DISCARDED

CATALOGING THE WORLD

CATALOGING THE WORLD

Paul Otlet and the Birth of the Information Age

Alex Wright

OXFORD
UNIVERSITY PRESS

Oxford University Press is a department of the
University of Oxford. It furthers the University's objective
of excellence in research, scholarship, and education
by publishing worldwide.

Oxford New York
Auckland Cape Town Dar es Salaam Hong Kong Karachi
Kuala Lumpur Madrid Melbourne Mexico City Nairobi
New Delhi Shanghai Taipei Toronto

With offices in
Argentina Austria Brazil Chile Czech Republic France Greece
Guatemala Hungary Italy Japan Poland Portugal Singapore
South Korea Switzerland Thailand Turkey Ukraine Vietnam

Published in the United States of America by
Oxford University Press
198 Madison Avenue, New York, NY 10016

Library of Congress Cataloging-in-Publication Data
Wright Alex, 1966–
Cataloging the world : Paul Otlet and the birth of the
information age / Alex Wright.
pages cm
Includes bibliographical references and index.
ISBN 978-0-19-993141-5 (acid-free paper) 1. Otlet, Paul, 1868–1944.
2. Bibliographers—Belgium—Biography.
3. Mundaneum—History. 4. Universal bibliography.
5. Documentation. 6. Classification—Books. 7. Information
organization—History. 8. World Wide Web—History. I. Title.
Z1004.O83W75 2014
020.9—dc23 2013035233

1 3 5 7 9 8 6 4 2

Printed in the United States of America
on acid-free paper

To Maaike

Arriving at each new city, the traveler finds again a past of his that he did not know he had.

—Italo Calvino

CONTENTS

CATALOGING THE WORLD

Introduction

When the Nazi library chief Hugo Krüss paid a visit to the Palais Mondial in Brussels's Parc du Cinquantenaire on December 4, 1940, he met a white-haired and frail-looking old man. The man's cheeks were hollowed, and his body seemed shrouded with age, but Krüss recognized him immediately as Paul Otlet.

The two men had met three years earlier, in Paris, at a scholarly conference. Then, they had come together as peers—Krüss, the powerful head of the Prussian State Library; and Otlet, the aging but still visionary information theorist and founder of the institute that organized the conference, where a parade of notable speakers (including Otlet, Krüss, and H.G. Wells) shared lofty visions of organizing the world's information.[1] Now, things had changed.

Krüss knew that Otlet had once been feted as a great man, keeping company with world leaders, Nobel laureates, and famous intellectuals. By 1940, however, he had lived long enough to see his fondest dreams undone, his reputation faded, and the country he loved conquered—twice. Now, he contented himself mostly with staying home, caring for his wife, remembering past glories, and nursing ancient vendettas. Nonetheless, faced with the representative of this second conquest, his views on certain subjects remained undimmed.

"He is a very old man with whom you can't talk about politics," wrote an unnamed Nazi apparatchik after an earlier meeting with Otlet on October 31, 1940. "He has peculiar fantasies about world peace."[2]

World peace was a distant dream. Europe was at war. Four months earlier, shortly after conquering Belgium on his way to France, Hitler had authorized a cultural task force to fan out across the occupied countries. Headed by his confidant Alfred Rosenberg—the recently appointed culture czar and author of a great deal of seminal and toxic Nazi philosophy—the group operated under orders to seize any and all valuable books, works of art, or religious objects from museums, libraries, and universities, as well as from private Jewish and Masonic collections. Drawing on this vast library of impounded books, Hitler hoped to build a new university worthy of the Third Reich, to be called the *Hohe Schule*.

As the long-serving head of Germany's greatest library, Krüss had risen to prominence in the Nazi political apparatus, earning the Führer's trust when he emerged as a highly visible apologist for the infamous Nazi book-burnings of 1933. By 1940 he had attained the rank of director general, and now served on Hitler's Privy Council. He had assumed a key role with the Rosenberg panel as one of its top three officials, and the leading authority on matters related to books, determining which titles to retain and which to discard from thousands of newly conquered libraries across Europe.[3] The commission would ultimately confiscate millions of volumes—including prized first editions, rare illuminated manuscripts, and countless other, more prosaic works on every topic imaginable—as well as paintings, religious objects, and innumerable other cultural artifacts. For all their rapaciousness, however, Krüss and his staff proved quite discriminating in choosing which books to keep. Most were destroyed or discarded. At one point the German army paved the muddy streets of Ukraine with thousands of seized books to speed the passage of their military vehicles. The Nazis had little use for novels, poetry, or devotional texts, but they were keenly interested in books containing practical scientific and technical information, as well as books about Judaism and Freemasonry.[4]

Those topics and many more they found amply represented in the Palais Mondial. For almost fifty years, Paul Otlet had worked tirelessly to build his Universal Bibliography, a vast index of published works that attempted to catalog every book, magazine, newspaper, and other significant piece of intellectual property ever created. At its peak, it numbered more than 15 million entries, all recorded on individual index cards and stored in a vast grid of wooden filing cabinets. In addition to the massive card catalog, the collection included more than 70,000 cardboard boxes' worth of documents, photographs, posters, pamphlets, and other documentary material, as well as a huge collection of museum pieces that had at various times filled up to 150 rooms in the Palais du Cinquantenaire (commissioned in 1880 by King Leopold II to celebrate Belgium's fiftieth anniversary of independence). As the collection grew, Otlet had developed a series of progressively more ambitious schemes to organize the collection and to promote universal access to human knowledge through a global information network that he dubbed the "Mundaneum." Now, the Nazis would determine the fate of his life's work.

The Rosenberg Commission's campaign of literary plunder, appalling as it was, hardly marked the first time a conquering nation had instituted a program of violent cultural appropriation. When King Ashurbanipal consolidated power over the Sumerian Empire in the seventh century BCE, he impounded every book in the kingdom to fill his royal library. Four hundred years later, the Egyptian pharaoh

Ptolemy I ordered his armed forces to search every incoming ship at the port of Alexandria and seize any books held onboard, thus populating the greatest library the world would know for another 2,000 years.

At other times, nations have simply chosen to destroy the intellectual heritage of the people they conquered. When Emperor Shi Huangdi consolidated power over the Chinese Empire in 213 BC, he commanded the destruction of every book in the kingdom to make way for a new library that better reflected his tastes. And when the Spanish conquistadors arrived in the Aztec kingdom in the fifteenth century, they promptly burned nearly all of the Aztecs' glyph-laden deerskin books—including treatises on law, mathematics, and herbalism (this destruction likely came as no great surprise to the Aztecs, who themselves had destroyed all the books of a previous conquered regime just a century earlier).

The fortunes of the great European powers have also ebbed and flowed along with their libraries. In 781, Charlemagne built a library at his court in Aachen, largely with books acquired from the Imperial Library at Constantinople. In the centuries that followed, generations of powerful popes would shore up the Vatican Library as they consolidated power across all of Christendom. More recent European empires—British, French, and Prussian among them—also erected enormous libraries commensurate with their imperial ambitions.[5]

A similar dream of cultural supremacy had led Krüss and his delegation to Paul Otlet. When they arrived at the Palais Mondial, they surveyed the collection that Otlet had assembled over the course of nearly half a century. In addition to the Universal Bibliography, there were exhibits on topics ranging from astronomy to paleontology to the history of Spain—encompassing works of art, photographs, textiles, and other material, much of it accompanied by explanatory diagrams and dioramas—as well as a seemingly endless parade of ephemera: old Belgian election posters, glass-plate photos of Egyptian archaeological digs, and instructions on how to make spectacles. As disparate as the collection might seem, it employed an extremely precise classification scheme known as the Universal Decimal Classification. Otlet had spent decades thinking about how to organize the descriptions and analyses of these items, and particularly how to synthesize and distribute them to a broader public. But the delegation saw little value in the apparent hodgepodge before them.

"The upper galleries . . . are one big pile of rubbish," one inspector noted in his report. "It is an impossible mess, and high time for this all to be cleared away." The Nazis evidently struggled to make sense of the curious spectacle before them. "The institute and its goals cannot be clearly defined. It is some sort of . . . 'museum for the whole world,' displayed through the most embarrassing and cheap and primitive methods." Nonetheless, the inspectors saw at least one thing of value in the collection: "The library is cobbled together and contains, besides a lot of waste, some things we can use. The card catalog might prove rather useful."[6]

Otlet interested the commission for another reason as well: his numerous foreign contacts. Over the years, he had formed a large network of friends and associates across the occupied nations, many of whom had similarly devoted themselves to the cause of world peace. The most prominent and long-standing of them was Henri La Fontaine, the Belgian politician, pacifist, and socialist who had won the Nobel Peace Prize in 1913. After World War I, the two men had even played a role in the formation of the League of Nations. That organization had failed to live up to Otlet's dreams as an international peacekeeping organization, however, failing utterly to prevent the German rise to power that now threatened to extinguish his most cherished dream.

What the Nazis saw as a "pile of rubbish," Otlet saw as the foundation for a global network that, one day, would make knowledge freely available to people all over the world. In 1934, he described his vision for a system of networked computers—"electric telescopes,"[7] he called them—that would allow people to search through millions of interlinked documents, images, and audio and video files. He imagined that individuals would have desktop workstations—each equipped with a viewing screen and multiple movable surfaces—connected to a central repository that would provide access to a wide range of resources on whatever topics might interest them. As the network spread, it would unite individuals and institutions of all stripes—from local bookstores and classrooms to universities and governments. The system would also feature so-called selection machines capable of pinpointing a particular passage or individual fact in a document stored on microfilm, retrieved via a mechanical indexing and retrieval tool.[8] He dubbed the whole thing a *réseau mondial*: a "worldwide network" or, as the scholar Charles van den Heuvel puts it, an "analog World Wide Web."[9]

Twenty-five years before the first microchip, forty years before the first personal computer, and fifty years before the first Web browser, Paul Otlet had envisioned something very much like today's Internet. In one remarkably prescient passage, he wrote:

> Everything in the universe, and everything of man, would be registered at a distance as it was produced. In this way a moving image of the world will be established, a true mirror of his memory. From a distance, everyone will be able to read text, enlarged and limited to the desired subject, projected on an individual screen. In this way, everyone from his armchair will be able to contemplate creation, in whole or in certain parts.[10]

Even more startling, Otlet also imagined that individuals would be able to upload files to central servers and communicate via wireless networks, anticipated the development of speech recognition tools, and described technologies for transmitting sense perceptions like taste and smell. He foresaw the possibilities of social networks, of letting users "participate, applaud, give ovations, sing in the chorus." And while he likely would have been flummoxed by the chaotic mesh of present-day social media outlets like Facebook and Twitter, nonetheless he saw the possibilities of constructing a social space around individual pieces of media, and allowing a network of contributors to create links from one to another, much the way hyperlinks work on today's Web.

The Mundaneum was to be more than just a networked library, however; Otlet envisioned it as a central component of a much vaster scheme to build a utopian World City. That city would sit at the center of a new world government, a transnational organization consisting of an international congress, judiciary, university, and a sprawling network of affiliated institutions and associations. An ardent "internationalist," Otlet believed in the inevitable progress of humanity toward a peaceful new future, in which the free flow of information over a distributed network would render traditional institutions—like state governments—anachronistic. Instead, he envisioned a dawning age of social progress, scientific achievement, and collective spiritual enlightenment. At the center of it all would stand the Mundaneum, a bulwark and beacon of truth for the whole world.

Otlet's expansive vision stood in stark contrast to the broken reality Hugo Krüss encountered that day in 1940. Compared with the stately museums and lavish art collections that he and the other inspectors were accustomed to looting, the Mundaneum must have seemed puzzling: less an august cultural institution than the clutter and sprawl of an eccentric old man. The collection had sat largely dormant for several years, ever since the Belgian government had withdrawn its funding for the project. But Otlet had soldiered on, curating the collection in private and writing about his dreams of a utopian, networked world.

Soon after the war started with the invasion of Poland in September 1939 and the subsequent British and French declarations of war, Otlet contemplated the looming terror that seemed to be unfolding. "This is the hour of great anguish. The two sides are almost upon each other," he wrote. "The hour when all of little Belgium may be for the second time embroiled in a war which is not its own. Horrors!"[11] Twenty-five years earlier, Otlet had lost his youngest son, Jean, at the Battle of Yser. Now, he faced the grim prospect of another German army massing across the eastern border.

On the eve of invasion, Otlet had sent a desperate telegram to President Franklin Roosevelt, imploring him to save the Mundaneum and offering the United States the entire collection "as nucleus of a great World Institution for World Peace and Progress with a seat in America."[12] The telegram even appeared in the Belgian press. Roosevelt sent no reply. After the invasion, he sent another telegram to Roosevelt but again received no response. Finally, in an apparent fit of desperation, he had even written a letter to Adolf Hitler, beseeching him to protect the Mundaneum.[13] He was prepared, it seemed, to try anything to save the project into which he had poured most of his life. Now, he threw himself on the mercy of the German inspectors. At the end of the visit, the inspector recounted that "Mr. Otlet made an appeal on behalf of humanity," pleading with the German delegation to preserve the collection he had spent the better part of five decades trying to assemble.

The Germans were not persuaded. Any such decision would have to be made by the Führer himself, they informed him. "Councilor

Krüss saw no chance of taking any immediate action, unless doing so would create space for defense or manufacturing purposes," wrote the official who chronicled the meeting.[14] Within a few days, a troop of German soldiers arrived and proceeded to clear out the contents of the Palais Mondial, eventually destroying sixty-three tons of books, journals, posters, pamphlets, and other documentary material that made up the core collection. Later that year, Nazi officials used the space formerly occupied by the Palais Mondial to host an exhibition of Third Reich art.

After the Nazis removed his material from the Palais Mondial, Otlet moved what remained of his collection to an unheated building near the Parc Léopold, where he continued to work quietly, tending to the surviving remnants of his collection. In the years that followed, he continued to work quietly in occupied Belgium, tending the dying embers of his dream, though still surfacing now and then to give an occasional lecture. He continued to develop his ideas about a global information network in private but increasingly found himself watching from a distance the machinations of a world that had already largely forgotten him.

On December 10, 1944, three months after the liberation of Brussels, Paul Otlet died. A few months later, on April 28, 1945, Hugo Krüss—recognizing the imminent German defeat—committed suicide in the basement of the Prussian State Library.

Over the years that followed, what remained of the Palais Mondial collection was gradually scattered around Brussels. Some boxes sat in an abandoned building in the Parc Léopold, eventually getting shunted around to other storage locations in the Chaussée de Louvain and later the Avenue Rogier.[15] At one point a large chunk of material found its way into an unlocked chamber adjacent to a Brussels subway station, where passers-by could reach in and help themselves to a stray book, journal, pamphlet, or photograph (the curators of the present-day Otlet archives occasionally have visitors stop in to return material they plucked from the subway years ago).[16]

The dismantling of the World Palace by the Nazis all but guaranteed the Mundaneum's plunge into historical obscurity. In the years following the war, most Belgians turned their attention to rebuilding their country and securing a better future. Outside the rarefied world of library and information science, almost no one in Belgium—or anywhere else, for that matter—had much cause to remember Paul Otlet. A small band of former disciples, the Friends of the World Palace ("Les Amis du Palais Mondial"), met annually to commemorate his work and lay garlands on a bust of him, commissioned after his death.[17]

In 1968, a young University of Chicago graduate student named Boyd Rayward made his way to Brussels to write a dissertation about Otlet, whose work he had encountered while pursuing his doctorate in library science. Following a few references in obscure library journals to Otlet's work, Rayward found himself increasingly entranced by the man, his work, and, as he later put it, "the riddles he presented."[18] At the time, Otlet seemed interesting primarily as an overlooked contributor to the practice of library cataloging; the World Wide Web was still more than twenty years away.

In that pre-digital era, conducting this kind of research required painstaking hours sifting through bibliographic references in musty journals—without the benefit of keyword searches—and combing the library stacks. Rayward's research eventually took him to Brussels, where he tracked down what remained of Otlet's collection in an abandoned office in the Parc Léopold, littered with books and manuscripts. The room he discovered was musty and strewn with cobwebs, with rainwater dripping from a leak in the ceiling. There was a faded garland wreath wrapped around his bust, placed long ago by the last gathering of the Friends.

It was something of a miracle that the cards and books—the remnants of Otlet's dream—had survived at all. In the years since World War II, very little effort had been made to reconstruct Otlet's collection, most of which had been scattered in 1940 by the Rosenberg Commission. The office was strewn with literary flotsam that appeared not to have been touched in decades: books, papers, manuscripts, and letters. Here Rayward discovered an old diary Otlet had kept dating back to his adolescence, as well as a large collection of archival papers and a few scribbled pages that Otlet had written to his second wife. But Rayward's time was limited; he had to fly back to Chicago soon, and he worried that if he didn't get what he needed on this trip, the deteriorating documents might not survive much longer (he was right; many of Otlet's diaries have since been lost).

As Rayward picked through the mounds of debris, he couldn't help but marvel at his subject's sheer determination and doggedness. In addition to the piles of notes and manuscripts, Otlet had amassed an enormous collection of newspapers, posters, postcards, photos, books, drawings, brochures, pamphlets, and all manner of other ephemera. The bookshelves groaned with cardboard file holders that looked ready to burst, their seams fraying with age. For all the seeming clutter, however, every item had its place. It seemed like the incarnation of Jorge Luis Borges's fictional Library of Babel, a theoretically infinite archive containing every possible combination of letters and numbers: "illuminated, solitary, infinite, perfectly motionless, equipped with precious volumes, useless, incorruptible, secret."[19]

In the decades that Otlet's papers had sat gathering dust, his dream of a universal knowledge network had found new expression across the

Atlantic, where a group of engineers and computer scientists laid the groundwork for what would eventually become the Internet. Beginning during the Cold War, the United States poured money into a series of advanced research projects that would eventually lead to the creation of the technologies underpinning the present-day Internet. In the 1990s, the World Wide Web appeared and quickly attracted a widespread audience, eventually establishing itself as the foundation of a global knowledge-sharing network much like the one that Otlet envisioned.

Today, the emergence of that network has triggered a series of dramatic—perhaps even "axial"[20]—transformations. In 2011, the world's population generated more than 1.8 zettabytes (1,800,000,000,000,000,000,000 bytes, or 1.8 trillion gigabytes) of data, including documents, images, phone calls, and radio and television signals.[21] More than a billion people now use Web browsers, and that number will almost certainly increase for years to come. In an era when almost anyone with a mobile phone can press a few keys to search the contents of the world's libraries, when millions of people negotiate their personal relationships via online social networks, and when institutions of all stripes find their operations disrupted by the sometimes wrenching effects of networks, it scarcely seems like hyperbole—and has even become cliché—to suggest that the advent of the Internet ranks as an event of epochal significance.

While Otlet did not by any stretch of the imagination "invent" the Internet—working as he did in an age before digital computers, magnetic storage, or packet-switching networks—nonetheless his vision looks nothing short of prophetic. In Otlet's day, microfilm may have qualified as the most advanced information storage technology, and the closest thing anyone had ever seen to a database was a drawer full of index cards. Yet despite these analog limitations, he envisioned a global network of interconnected institutions that would alter the flow of information around the world, and in the process lead to profound social, cultural, and political transformations.

By today's standards, Otlet's proto-Web was a clumsy affair, relying on a patchwork system of index cards, file cabinets, telegraph machines, and a small army of clerical workers. But in his writing he looked far ahead to a future in which networks circled the globe and data could travel freely. Moreover, he imagined a wide range of expression taking shape across the network: distributed encyclopedias, virtual classrooms, three-dimensional information spaces, social networks, and other forms of knowledge that anticipated the hyperlinked structure of today's Web. He saw these developments as fundamentally connected to a larger utopian project that would bring the world closer to a state of permanent and lasting peace and toward a state of collective spiritual enlightenment.

The conventional history of the Internet traces its roots through an Anglo-American lineage of early computer scientists like Charles Babbage, Ada Lovelace, and Alan Turing; networking visionaries like Vinton G. Cerf and Robert E. Kahn; as well as hypertext seers like Vannevar Bush, J. C. R. Licklider, Douglas Engelbart, Ted Nelson, and of course Tim Berners-Lee and Robert Cailliau, who in 1991 released their first version of the World Wide Web. The dominant influence of the modern computer industry has placed computer science at the center of this story.

Nonetheless Otlet's work, grounded in an age before microchips and semiconductors, opened the door to an alternative stream of thought, one undergirding our present-day information age even though it has little to do with the history of digital computing. Well before the first Web servers started sending data packets across the Internet, a number of other early twentieth-century figures were pondering the possibility of a new, networked society: H. G. Wells, the English science fiction writer and social activist, who dreamed of building a World Brain; Emanuel Goldberg, a Russian Jew who invented a fully functional mechanical search engine in 1930s Germany before fleeing the Nazis; Scotland's Patrick Geddes and Austria's Otto Neurath, who both explored new kinds of highly designed, propagandistic museum exhibits designed to foster social change; Germany's Wilhelm Ostwald, the Nobel Prize–winning chemist who aspired to build a vast new "brain of humanity"; the sculptor Hendrik Andersen and the architect Le Corbusier, both of whom dreamed of designing a World City to house a new, one-world government with a networked information repository at its epicenter. Each shared a commitment to social transformation through the use of available technologies. They also each shared a direct connection to Paul Otlet, who seems to connect a series of major turning points in the history of the early twentieth-century information age, synthesizing and incorporating their ideas along with his own, and ultimately coming tantalizingly close to building a fully integrated global information network.

Despite the occasional newspaper article about Otlet's work as a conceptual forerunner to the Web,[22] the larger contours of his story remain little known outside of specialized academic circles. For anyone who follows the technology industry, this should come as no surprise. Computer scientists often show little enough interest in their own history, let alone the contributions of a group of long-dead Europeans (none of them programmers), most of whom never published in English. As a result, their story has remained relegated to the historical margins.

Otlet not only invites study as an early avatar of the networked age. His life and work also shed light on the deeper causes and conditions of the information age in which we now live. While the proliferation of computers in recent years has certainly contributed to the much-chronicled problem of information overload, the first rumblings of our present-day data deluge really started during the second Industrial Revolution of the late nineteenth and early twentieth centuries, when the citizens of Europe and North America experienced a series of unprecedented technological transformations. In the span of just a few decades, an enormous number of innovations were unleashed: automobiles, airplanes, radio, telegraphs, typewriters, punch cards, microfilm, vaccines, and mechanical weapons. Taken together, these rapid advances created a state of technological culture shock. People and nations that had once lived in relative isolation quickly found themselves intertwined in a mesh of complex networks. Telegraphs, telephones, railways, post offices, and expanding road systems allowed people, goods, and money to move across national borders more easily than ever before. As a result, ideas started moving more freely as well, triggering an explosion of published information. The resulting intellectual, commercial, and political entanglements gave rise to whole new industries, professions, and modes of thought, as well as to new opportunities for conflict—reaching the horrific crescendo of World War I. This gave rise to a new "internationalist" consciousness that inspired some to imagine fundamentally new modes of living in a global, networked society.

Otlet embraced the new internationalism and emerged as one of its most prominent apostles in Europe in the early twentieth century. In his work we can see many of these trends intersecting—the rise of industrial technologies, the problem of managing humanity's growing intellectual output, and the birth of a new internationalism. To sustain it Otlet tried to assemble a great catalog of the world's published information, create an encyclopedic atlas of human knowledge, build a network of federated museums and other cultural institutions, and establish a World City that would serve as the headquarters for a new world government. For Otlet these were not disconnected activities but part of a larger vision of worldwide harmony. In his later years he started to describe the Mundaneum in transcendental terms, envisioning his global knowledge network as something akin to a universal consciousness and as a gateway to collective enlightenment.

The Mundaneum was to be more than just an information-retrieval tool. It would form an essential component of a much larger scheme to unite the world's nations under a new form of government, leading to a new age of peace and understanding, one in which the traditional warring factions of nation-states and calcifying political structures would give way to a networked world. In such an environment, he believed, humanity could finally reach its true spiritual potential.

By the time of the Nazi occupation of Belgium, however, such visions seemed quaintly passé. By the end of the 1930s he already seemed a relic from an earlier era—an era when Europeans dreamed of harmony during the belle epoque, when the wonders of industrial technology and the flourishing of published knowledge seemed to hold enormous potential. Those dreams of technological progress were blunted by the brutal reality of mechanized warfare during World War I. And now, on the eve of the second World War, the dark side of technological progress was coming horrifically into focus. The information technologies that would follow in the wake of the war—including modern computers and the Internet—were developed not out of spiritual idealism but largely at the behest of military institutions. But the vision of Otlet and others created an intellectual and—not to be too metaphysical—spiritual context in which these inventions could take shape.

Much of the techno-dystopian rhetoric that has surrounded the rise of the Web over the past few years seems predicated on the notion that we are living through a unique period of technological disruption. But some, such as the historian Peter Watson, have argued that compared to the late nineteenth century we may actually be living through a period of relative technological calm. This is not to diminish the importance or usefulness of our digital-age inventions; it is rather to put them in context of a period of even more fundamental technological disruption stretching back well into the late nineteenth century. While Paul Otlet and his contemporaries—artists, writers, engineers, dreamers—may not have played a direct role in the development of the Internet, their work illuminates the deeper historical forces behind it. They open a window into the early twentieth century, helping us understand the social, economic, and technological forces that gave rise to the networked society in which we all live.

During its brief heyday, Otlet's Mundaneum was also a window onto the world ahead: a vision of a networked information system spanning the globe. Today's Internet represents both a manifestation of Otlet's dream and also, arguably, the realization of his worst fears. For the system he imagined differed in crucial ways from the global computer network that would ultimately take shape during the Cold War. He must have sensed that his dream was over when he confronted Krüss and the Nazi delegation on that day in 1940. But before we can fully grasp the importance of Otlet's vision, we need to look further back, to where it all began.

1

The Libraries of Babel

In the mid-sixteenth century, a Swiss naturalist named Conrad Gessner hit upon a novel method for organizing his notes. "Cut out everything you have copied with a pair of scissors," he wrote, then "arrange the slips as you desire."[1] Using this technique, Gessner culled material from a wide range of sources—personal notes and observations, passages from books (cutting them out rather than copying them by hand, as a time-saving technique), recipes, printers' catalogs, letters from friends, and so on—then collected the contents into a series of works that he published as new books.[2] The most famous of these was his *Bibliotheca Universalis*, a prodigious volume encompassing some 10,000 titles by 3,000 different authors and purporting to catalog every known book.[3]

Gessner's collection may seem modest by contemporary standards—the Library of Congress now adds about 10,000 items to its collection every day[4]—but in 1545 that number constituted an appreciable chunk of the available supply of recorded knowledge. At the time, the Vatican Library—then the largest library in all of Europe—housed fewer than 5,000 volumes.

Gessner was scarcely alone in his urge to tackle the problem of information overload. In the years following Gutenberg's invention of the printing press, the volume of printed material grew at

BIBLIOTHECA

Vniuerſalis, ſiue Catalogus omni- um ſcriptorum locupletiſſimus, in tribus linguis, Latina, Græca, & Hebraica: extantium & non extantiū, ueterum & recentiorum in hunc uſq; diem, doctorum & indoctorum, publicatorum & in Bibliothecis latentium. Opus nouum, & nō Bibliothecis tantum publicis priuatiſue inſtituendis neceſſarium, ſed ſtudioſis omnibus cuiuſcunq; artis aut ſcientiæ ad ſtudia melius formanda utiliſſimum: authore CONRADO GESNERO Tigurino doctore medico.

TIGVRI APVD CHRISTOPHORVM Froſchouerum Menſe Septembri, Anno M. D. XLV.

Conrad Gessner's *Bibliotheca Universalis* (1545).

an astonishing clip. By 1500, just fifty years after the introduction of movable type, European printers had turned out an estimated 20 million copies of between 15,000 and 20,000 individual titles. A century later, that number would grow to an estimated 200 million.[5] This early modern information explosion did not happen solely as the result of technological innovation. The historian Ann Blair, for one, has argued that the proliferation of printed books was only one of several forces driving the unprecedented outpouring of published knowledge in the fifteenth and sixteenth centuries. The discovery of new worlds, the recovery of ancient texts from the Greeks, and the rise of secular universities all contributed to a growing demand for new sources of knowledge.[6] These developments, coupled with the invention of the powerful new technology of printing, created a fast-growing marketplace for ideas that would continue to accelerate in the centuries ahead.

Whereas books had once been largely the province of church, state, and university—and mostly the church—now they started to fall into the hands of private citizens, although they remained quite expensive by contemporary standards. A typical seventeenth-century English household might own a single book, such as a devotional Book of Hours or John Bunyan's *Pilgrim's Progress,* but a growing class of scholars, merchants, and families of means began building personal libraries. Meanwhile, the rising popular demand for books fueled the spread of open-air book markets in major European cities. As books escaped the confines of medieval scriptoria and university libraries, a new class of secular writers began to emerge: humanists, scientists (or "natural philosophers," as they called themselves), poets, fabulists, and encyclopedists.

Responding to this outpouring, several European scholars began experimenting with ideas for synthesizing the growing spectrum of written thought. Between 1500 and 1700, some European publishers started producing books that today we would call reference works: encyclopedias, dictionaries, collections of quotations and historical anecdotes, and other "books about books," like bibliographies, and library and bookseller catalogs. Many came outfitted with new tools to help readers locate information stored between their covers. Medieval scribes had often used highly idiosyncratic methods to organize the information in their collections—visualizations, rhyming, or other bespoke methods. Secular publishers recognized that readers often wanted to locate information quickly, without having to master a new mnemonic system for each book. Thus they began to experiment with standardized, easy-to-use devices like tables of contents, outline and page-layout conventions, and alphabetic indexes.[7]

The rising demand for books created strong incentives for publishers to streamline their production processes. Compilers of reference books would often cut and paste material directly from one text to another as a time-saving measure, sometimes even importing complete chunks of text (belonging to themselves or others). Untold thousands of medieval manuscripts were butchered in this way, as printers plucked their contents and rearranged them into new books, plundered illuminations, or salvaged parchment or binding to make new books. Early Renaissance publishers often saw older books as little more than raw material, ready to be reused and remixed into new forms: the original literary mash-up.

The apogee of early European reference works may have been the London polyglot Bible of 1657, which presented parallel versions of the scriptures in Greek, Latin, Hebrew, Samaritan, Greek, Chaldee, Syriac, Arabic, Persian, and Ethiopian. The text also included a lengthy set of appendices that included a chronology of the Bible, maps of the Holy Land, and plans of the temple; essays on coins, weights, and measures, as well as the origin of languages;

a history of different versions of the scriptures; and a table of variant readings.[8]

So prevalent did reference books become during the Renaissance that they gradually found their way into popular lore—though not always in the most positive light. In Christopher Marlowe's famous telling of the Faustus story, the doctor strikes his bargain with the devil Mephistopheles in exchange for a kind of encyclopedia (a *Thesaurus Naturae,* or *Natural Treasury*). Ignoring the Good Angel's plea to "lay that damned book aside/and gaze not upon it, lest it tempt thy soul," Faustus takes the magical book in hand and consigns himself to perdition. The prize that costs Faustus his soul is, paradoxically, a short book—one that due to the black magic of printing managed to encompass the entirety of human and divine knowledge in a single slim volume.

Faustus's fatal fascination with the encyclopedia reflected a widespread popular distrust of books and bookmaking that held sway throughout Western Europe during the period. In the years following Gutenberg, as printing presses started cropping up everywhere from Mainz to Venice to London, ordinary Europeans began to regard books with an equal sense of wonder and suspicion. Renaissance folklore held that a devil inhabited every print shop—which may explain why printers often dubbed their apprentices "printer's devils." The association of printing with the diabolical arts may well stretch back to the very earliest days of the printing press. Johannes Gutenberg's first business partner was a fellow named Johann Fust, who eventually faced charges of witchcraft after showing some of his newly printed Bibles to the Archbishop of Paris (forcing him to reveal his secret). In the decades to come, Fust's name would become conflated with the similar-sounding Doctor Faustus, and the history of the printing press would thus intermingle with the cautionary fable of a man who lusted after too much knowledge. Thus for many years to come, the spread of printing carried with it the intimations of a devil's grand bargain.

European skepticism about the benefits of books may have had deeper social roots as well. Throughout most of the history of the written word, only a small class of rulers and scholars knew how to read; for centuries, books had remained the province of the powerful and moneyed elites. In an age when most people still remained illiterate, then, perhaps it should come as no surprise that the literate elites might view the printing press with suspicion. Such an aversion to printed books may seem difficult for many of us to fathom today. After all, most of us grow up in a culture that values literacy and learning as unqualifiedly Good Things. But it was not always so. In the Middle Ages, many considered intellectual curiosity a devilish impulse. In the monastic scriptoria—where monks worked tirelessly to preserve the heritage of early Christian and classical writing—the monasteries often discouraged scholarly inquiry in favor of quiet devotion and contemplation. The influential Irish monk Columbanus even issued a rule forbidding monks to disagree with each other, decreeing the offense punishable by fifty blows.[9]

The classical pursuit of knowledge that had so animated the Greek and Roman philosophers found little purchase in the rigid intellectual strictures of the medieval church; indeed, later Renaissance scholars regularly castigated medieval monks for their plodding devotion and lack of intellectual risk-taking. Even after Gutenberg's invention gave rise to a broader secular literature, however, many prominent thinkers continued to question whether the proliferation of published knowledge was such a good thing after all. "Is there anywhere on earth exempt from these swarms of new books?" asked Erasmus in 1525. "Even if taken out, one at a time, they offered something worth knowing, the very mass of them would be a serious impediment to knowing."[10] Elsewhere, John

Calvin complained of "that confused forest of books."[11] Even Gessner, who had found his calling in compiling books drawn from the works of others, bemoaned the "silliness of useless writings in our time" and "the harmful and confusing abundance of books."[12] Less alarmist writers of the Renaissance objected to the proliferation of books not on diabolical grounds but purely for practical reasons: There were simply more books than anyone could ever hope to read. "Even if all knowledge could be found in books, where it is mixed in with so many useless things and confusingly heaped in such large volumes," wrote René Descartes, "it would take longer to read those books than we have to live in this life."[13]

Some scholars, however, celebrated the benefactions of the printing press. The English historian John Foxe called it a "divine and miraculous invention," giving voice to a strain of technological optimism not far removed from the utopian rhetoric of today's Internet acolytes. "Hereby tongues are known, knowledge grows, judgment increases, books are dispersed, the scripture is seen, the doctors are read, histories opened, times compared, truth discerned, falsehood detected, and all (as I said) through the benefit of printing."[14] Foxe benefited enormously from the printing press himself, writing a comprehensive history of England that sported a rather imposing title: *The First Volume of the Ecclesiastical History Containing the Acts and Monuments of Things Passed in Every King's Time in this Realm, Especially in the Church of England Principally to be Noted, with a Full Discourse of Each Persecutions, Horrible Troubles, the Suffering of Martyrs, and Other Things Incident Touching As Well the Said Church of England as also Scotland, and All Other Foreign Nations, from the Primitive Time Til the Reign of King Henry VIII* (more commonly known, to the great relief of historians, as *The Acts and Monuments*). The contents of the book proved the exhaustive equal of its title, with more than 800 pages comprising a thorough retelling of English history drawn from a wide range of earlier published sources. Other English historians began publishing similarly elaborate encyclopedic histories, including John Stow's *Annals of England* (1580) and Ralph Holinshed's *Chronicles of England*, which synthesized previous works. These were vastly ambitious projects, brimming with details borrowed liberally from other sources. These new synthetic histories seemed to strike a particular chord with the reading public, which seemed eager for books that aspired toward completeness about a given topic.[15]

As more and more of these encyclopedic books appeared across Europe, a few scholars began exploring more sophisticated methods for synthesizing material from disparate sources. In 1627, Francis Bacon published his *Sylva Sylvarum; or, A Natural History*, offering an example of how knowledge could advance through the interpolation of previously published works. Here he chronicled the results of 1,000 experiments, collected from both his own work and the published reports of other scholars. Bacon also thought deeply about how to classify the growing accumulation of published information. Taking Aristotle's classification scheme as a reference point, he proposed a new and, he argued, much-improved system that situated all human learning (as distinct from divine learning) in terms of three basic areas of inquiry: history, poetry, and philosophy. He expounded on this classification in detail in his 1623 work *De Augmentis Scientiarum; or, The Arrangement and General Survey of Knowledge*. Whereas Aristotle's system had focused primarily on the classification of things—plants, animals, persons, and other relatively concrete entities—Bacon aimed to classify knowledge itself: not things, but ideas.

Bacon divided his scheme into two top-level categories: divine learning, encompassing the timeless truth of the scriptures, and human learning, encompassing all forms of mortal knowledge,

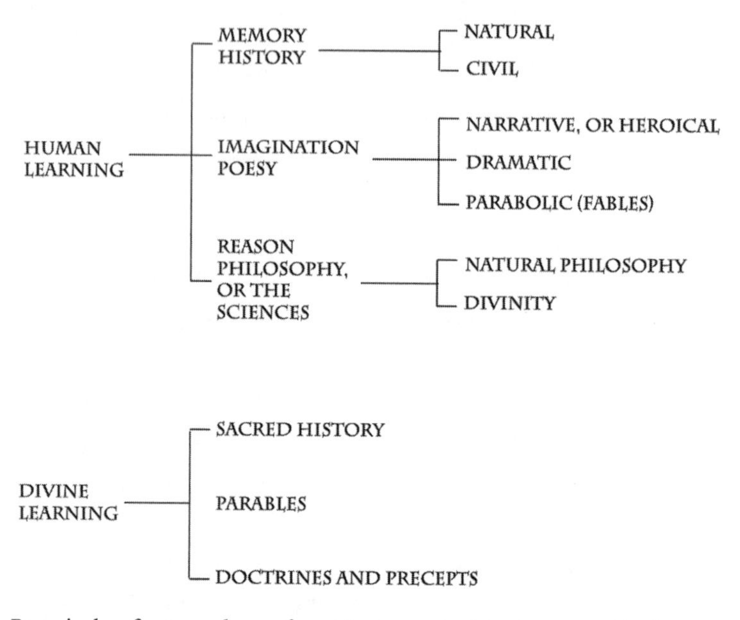

Bacon's classification scheme, from *De Augmentis Scientiarum; or, The Arrangement and General Survey of Knowledge* (1623). Illustration by Alex Wright.

including history, poetry, and philosophy (then a broad concept encompassing all of what we would today call the sciences). As proscriptive as such a categorization might seem, Bacon implored his readers not to interpret it as a rigid separation of the disciplines. He hoped, instead, "that all partitions of knowledge be accepted rather for lines and veins, than for sections and separations; and that the continuance and entireness of knowledge be preserved."[16] In other words: That his classification would foster the unity, rather than the atomization, of human knowledge.

That sentiment seems eerily prescient, insofar as it anticipated one of the central tensions that would bedevil the academic world in generations to come: the increasing separation of scholarly disciplines.

Bacon hoped to develop a universal encyclopedia based on his classification system that would encompass the entirety of human thought. He recognized, however, that such an undertaking would need more than a single author. At one point he even pressed King James I to lend state support to the project. But it would take another century for his vision to come to fruition under the guidance of the French encyclopedist Denis Diderot, who attributed the inspiration for his landmark encyclopedia to Bacon. "If we come out successful from this vast undertaking," he wrote, "we shall owe it mainly to Chancellor Bacon, who sketched the plan of a universal dictionary of sciences and arts at a time when there were not, so to speak, either arts or sciences."[17]

Bacon's philosophical framework would reverberate for centuries, influencing later generations of thinkers whom we might today call information scientists, including not only Diderot but Ephraim Chambers and Thomas Jefferson, who used Bacon's scheme as the conceptual basis for his own personal library catalog, which would go on to serve as the foundation of the Library of Congress. The American librarian Melvil Dewey would later adapt Bacon's scheme for his Decimal System, which in turn provided the conceptual foundation for another cataloging scheme that would take shape in Europe at the beginning of the twentieth century: the Universal Decimal Classification, invented by Paul Otlet.

In 1668, the English bishop John Wilkins published *An Essay Towards a Real Character and a Philosophical Language*, in which he unveiled his proposal for a new universal language. He described a kind of synthetic lingua franca incorporating a sweeping classification of all human knowledge, which he hoped would one day become "legible by any Nation in their own Tongue." Intended to resemble the Great Chain of Being proposed by Aristotle—revealing the hierarchy of nature, from highest forms to lowest—Wilkins's scheme divided the entirety of the world's accumulated knowledge into two broad categories: "General or universal notions" and "Special" topics. These categories were subdivided into 40 separate classifications, which were subsequently classified into a further 4,194 subcategories. These headings were then linked by a grammar of additional conjugations, suffixes, and other syntactic elements that would allow the speaker to link one thought to another. He also proposed the use of simple signs, like dashes and other symbols, to express differences or opposition between concepts. (an innovation that Otlet too would eventually embrace). The eminent semiotician Umberto Eco describes Wilkins's scheme as "a system of transcendental particles."[18] For a time, Wilkins's system took hold in the library of the Royal Society.

Bacon's and Wilkins's systems provided their intellectual descendants with new theoretical foundations for thinking about how to organize knowledge. Over the centuries that followed, their work wound its way into the intellectual fabric of European scholarship. Bacon's approach to "natural philosophy" influenced generations of scientists, subtly shaping the trajectory of research through his delineation of the disciplines. There is also a direct ancestral lineage from Bacon's seventeenth-century schema to the organization of modern libraries. Wilkins's work also influenced generations of thinkers from the Enlightenment forward. And while his universal language never made its way into common parlance, he did propose another idea that turned out to have legs: the metric system.

As the volume of published information continued to escalate, scholars and publishers began casting around for technical solutions to manage the growing data deluge. As the years passed, Gessner's humble paper-slip method proved a remarkably durable technology. In 1676, the German philosopher and mathematician Gottfried Wilhelm Leibniz drew on Gessner's approach as well, using a so-called excerpt cabinet (*scrinium literatum*) that allowed him to catalog his own prodigious collection of paper slips, which he used, sometimes obsessively, to document almost every thought that ever occurred to him while reading books, taking long walks, or traveling around. Devising the cabinet, Leibniz drew on his considerable skills as a tinkerer and inventor. In years past he had also invented a number of calculating machines, including one called the Stepped Reckoner—the first mechanical device capable of performing basic mathematical functions: addition, subtraction, multiplication, and division. During his years as a Jesuit student, Leibniz had learned of a Chinese numerical system that had received scant attention in Europe: the binary system. He began to ponder the possibilities of the binary scale in theological terms, wondering whether "God, represented by the number 1, created the Universe out of nothing, represented by 0."[19] Some historians have argued that Leibniz's work paved the way for the modern digital computer.[20]

A polymath who cast his prodigious mind over a wide range of topics—from infinitesimal calculus to Confucianism, theodicy, and metaphysics—Leibniz too worried about the explosion of published knowledge. Fretting over "that horrible mass of books which keeps on growing," he warned that "the indefinite multitude of authors will shortly expose them all to the danger of general oblivion" and, ultimately, "a return to barbarism."[21] Leibniz experienced that horrible mass more keenly than most. For a time he worked as the librarian at the library of Wolfenbuettel, a town in lower Saxony, which housed more than 100,000 volumes—making it one of Europe's largest libraries at the time. He designed both a building for the library and a new cataloging system to guide the arrangement of books. At one point he tried to launch a campaign to ask publishers to send him brief summaries of each new title, so that he could enter them into a central, searchable index.

Leibniz's efforts at creating a new library cataloging and indexing system met with mixed success. But his excerpt cabinet proved a highly effective personal solution to the problem of information overload. Coming home at the end of the day, he would sort through his own slips of paper and arrange them into the specially made cabinet, which featured small labels affixed to each shelf to identify the topic contained thereon.

The paper slip continued to find new adherents in the years that followed. In the mid-eighteenth century, Samuel Johnson used a variation of Gessner's technique to produce his famous dictionary, the first of the English language, employing a team of assistants to cut up his drafts onto paper slips and sort them into alphabetical order, then glue them into a master manuscript.[22] A century later, the *Oxford English Dictionary* would follow a similar scheme, as its instigator, James Murray, supervised the accumulation of millions of paper slips stored in 1,029 pigeonholes. In the late eighteenth

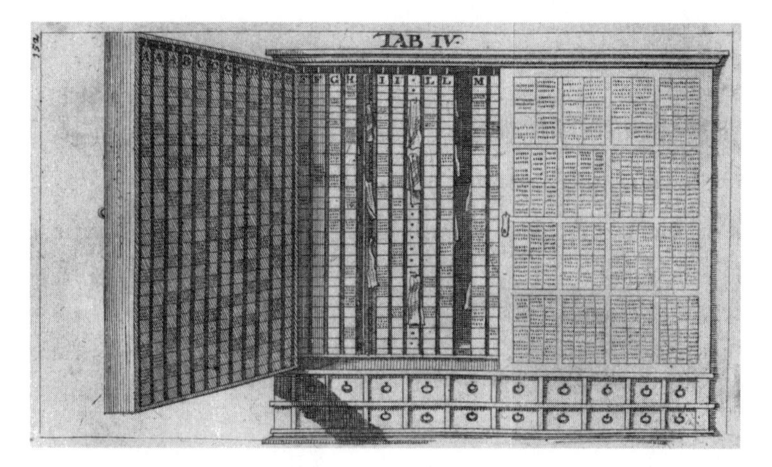

Excerpt cabinet, from Vincent Placcius's *De Arte Excerpendi* (1689). Reproduced with permission from the Houghton Library, Harvard University. GC6. P6904.689d, tabula IV and V after p. 153.

century, a Frenchman named Abbé François Rozier took on the formidable task of cataloging the library of the French Academy of Science. Taking Gessner's method a step further, he not only separated his notes into individual entries, but he also settled on a more durable form of data storage: playing cards. Unlike their modern counterparts, eighteenth-century playing cards featured no decorative patterns on their backs. These blank white surfaces made them ideal for use as note cards, lottery tickets, marriage and death announcements, and business cards. Playing cards also had another virtue: Most of them adhered to a standard size (either 83 x 43 mm or 70 x 43 mm). Stiff, evenly sized, and easy to flip with one hand, playing cards thus made the ideal building blocks for a library catalog.

In 1780, an Austrian named Gerhard van Swieten further adapted the technique to create a master catalog for the Austrian National Library, known as the Josephinian Catalog (named for Austria's "enlightened despot" Joseph II). Van Swieten decided to store his catalog cards in 205 wooden boxes, sealed in an airtight locker—the first recognizable precursor to the once familiar, now rapidly disappearing, library card catalog.[23]

Today, we might tend to think of the card catalog as a simplistic information retrieval tool: the dominion of somber librarians in fusty reading rooms. However, to take such a dismissive view of these compact, efficient systems—the direct ancestors of the modern database—may lead us to overlook the critical role they played in the industrial information explosion that would reshape the European world in the nineteenth century.

In 1832, William Strange published the first volume of his *Penny Story-Teller*, a forty-six-page chapbook printed on cheap paper whose table of contents listed such fare as "The Cure of Consumption,"

"House and Household," and "Conscience Makes Cowards." While the magazine may not have left much of a mark on the literary world, it signaled the dawn of a new era in publishing.

By the early nineteenth century, the process of industrialization had already started to transform England's economy, driving huge numbers of residents from the villages to the growing industrial cities, accompanied by the spread of a public school system that was rapidly increasing the size of the literate populace. By 1832, fully two-thirds of factory workers could read. Meanwhile, the new industrial printing technology—like steam-powered engines—made it possible to produce cheap, mass-produced literature that turned a steady profit. Strange's one-penny pamphlet inaugurated this new literary movement, which was literally steam-powered. The imitators soon came along, turning out serialized popular fiction in the form of so-called Penny Bloods or Penny Dreadfuls (one of which, *The String of Pearls*, introduced in 1846 that most dreadful of characters, Sweeney Todd). Stories of pirates and highwaymen found a ready audience among the newly literate urban readers. Soon thereafter followed popular novels, posters, postcards, and vast numbers of little pamphlets on topics like trade unions, socialism, and woman suffrage. Eventually, Charles Dickens would reap the benefits of this mass audience as he started producing the Pickwick Papers.

Mass-produced popular literature only tells a part of the story of the industrial information explosion. While the first wave of industrialization had transformed the production of textiles with iron and steam engines, by the second half of the nineteenth century a so-called second Industrial Revolution had started to take shape. The invention of electricity and the internal combustion engine accelerated the production and transportation of goods, while new communications technologies like the telegraph and the radio started to lay the foundations for a nascent global communications network.

These developments in turn created the conditions for new patterns of knowledge production to emerge.

As industrialization took hold across England and eventually the rest of Europe, so too did a new organizational ethos. The principles of scientific management, as it was called, most famously articulated by the American mechanical engineer Frederick Winslow Taylor, began to shape the work practices of many growing industrial companies. Giving eloquent voice to the new industrial ethos, Taylor described how professional managers must learn to take a whole-systems view of their organizations, instituting tightly controlled processes to maximize productivity. "In the past the man has been first; in the future the system must be first."[24] Using time-motion studies, log books, and other observational methods, Taylor developed a framework for optimizing the manufacturing process by creating standardized operating procedures and rigorous measurements of employee output. Thus was born the factory assembly line, where each worker was recast as, in effect, part of the machine. But the Taylorist ethos extended beyond the factory floor, shaping the roles of managers and other professional staff, and ultimately extending into the realm of nonmanufacturing organizations as well. Scholar Michael Buckland describes the Taylorist spirit as one of "technological modernism," rooted in an unshakeable faith in the liberating power of industrial progress. "The belief was that technology plus standards plus systems would induce progress," Buckland writes, "progress toward liberal ideas for some, progress towards competitive advantage for others."[25]

Today, in our postindustrial era, we may harbor a reflexive tendency to regard the process of industrialization through the prism of its less savory consequences: the disruption of traditional ways of life, the environmental costs, and the assault on individual dignity conjured in William Blake's "dark Satanic Mills." The enduring critiques of writers like Dickens and William Morris have done much to lade the word "industrialization" with strong negative connotations. However, at the time, most learned Europeans viewed industrialization as a boon. The Enlightenment ideals of scientific progress and social liberalization helped propel a widespread belief in the possibility of human advancement and moral uplift through technological innovation. Taylorist scientific management principles pointed not just toward greater economic output and organizational efficiency, but toward a new era of human potential in which wealth would be created and shared, old class lines would give way to a meritocracy of achievement, and eventually a utopian society might finally emerge.

Scientific management techniques would eventually reshape not just the production of goods and services, but also the contours of human knowledge itself. As companies, governments, universities, and other institutions expanded throughout the nineteenth century, the people who ran those organizations also began producing more and more paper—not just books and journals. Abetted by new information technologies like the typewriter, file folders, filing cabinets, and even paper clips, they wrote letters, memos, reports, white papers, pamphlets, scientific papers, operations manuals, train schedules, inventories, and all manner of other documents. Archivists began taking advantage of new technologies—particularly those paper clips and file folders—to manage the deluge. While much of this material was intended for internal consumption, many of these documents—especially those produced by governments and nonprofit institutions—were also intended for public consumption, making their way into bookstores and libraries and circulating via an increasingly widespread international postal system. By the end of the nineteenth century, the world was awash in paper.

The problem of cataloging so much information first came to a head, naturally, in the library world, where administrators started casting around for new methods to cope with the flood of documents. As Taylor's efficiency doctrines created the industrial conditions for the information explosion of the late nineteenth century, the problem of information management began to capture the attention of energetic librarians like the aforementioned Melvil Dewey, the great proselytizer of card cataloging whose eponymous Decimal System continues to serve as the ontological foundation for library catalogs across North America and elsewhere. Along with his less well-known contemporary Charles Cutter—whose pioneering work at the Harvard College library and, later, the Boston Athenaeum would provide the conceptual foundations for the Library of Congress catalog—Dewey started to develop new strategies for libraries that would enable them to cope with the barrage of documents demanding their attention.

An "irrepressible reformer" (as his biographer, Wayne A. Wiegand, calls him), Dewey devoted most of his life to pressing for social change: railing against alcohol and tobacco, promoting the metric system, and even agitating to simplify English spelling (even going so far as to change the spelling of his own name to the phonetically correct Melvil Dui). Dewey invented his Decimal Classification while still an undergraduate at Amherst College, drawing on the earlier work of important library thinkers like Cutter and William T. Harris, whose cataloging scheme for the St. Louis Library had drawn directly on the concept of Francis Bacon and his division of all learning into three high-level categories: history, poetry, and philosophy.[26] Soon after submitting his original scheme to the Amherst College Library Committee shortly before his graduation in 1873, he accepted a position as an assistant librarian there and set about trying to implement his ideas. By 1875 he had taken charge of the library. The next year, he collaborated with Cutter to establish the American Library Association, the entity that would ultimately give Dewey a vehicle for driving his agenda of institutional reform across the American library world. That role ultimately led him to become chief librarian at Columbia University, where he continued to develop his ideas, eventually establishing the nation's first professional library school.

Dewey subscribed wholeheartedly to the doctrines of industrial management, arguing that in order to achieve their collective potential as vessels for public education and self-improvement, libraries would have to dispense with "wasteful and unsatisfactory" methods and accept a high degree of standardization that would make them more efficient.[27] When he formally unveiled his cataloging system to the public in 1876, he tried to attract interest by touting its economic benefits, promising an "increased usefulness" that could be realized "more economically than by any other method."[28] For libraries to thrive in an industrial age, Dewey argued, they would have to abandon their idiosyncratic local conventions and adopt standardized techniques that would eventually yield massive economies of scale. By sacrificing individual autonomy in the service of the greater ideal of productivity, Dewey's system brought Taylorism out of the factory and into the library stacks. To realize that potential, Dewey insisted that libraries would need to submit themselves to a program of strict standardization: management techniques, equipment, and systematic rules that would govern everything from the printing of index cards to the size of drawers, boxes, inkwells, and pens—all in the hope of achieving a greater level of synchronization. The Dewey system offered librarians a standard organizing scheme that allowed them to acquire and catalog material efficiently and share bibliographic information with other institutions. Dewey saw these operational improvements as part of a larger social mission. By helping libraries expand their collections and make their material more widely available, they would serve more patrons and thereby exert a growing influence over society at large.

Dewey's classification serves as a case study in industrial management techniques. It rests on two conceptual foundations: first, a tightly controlled vocabulary of subject headings; second, an artificial notation that relies on numbers, letters, and other symbols to organize books into nine top-level "classes," each with a corresponding beginning number. Each class is further subdivided into ten "divisions," which in turn are subdivided into 1,000 distinct headings. It started with a broad category for Generalities (000s); Philosophy (100s) came next, because the study of philosophy ultimately provided the basis on which all other forms of knowledge rested. Theology (200s) naturally followed out of Philosophy, because religion is the means by which philosophy acquires its authority. Then came the social sciences, which then came under the rubric Sociology (300s), followed by language, then called Philology (400s), Natural Science (500s), technology, then known as the Useful Arts (600s), and Fine Arts (700). Finally, at the higher levels of cultural abstraction came Literature (800) and History (900). The classification has evolved slightly over time but retains its basic contours.

Dewey's system has drawn criticism over the years for its seeming arbitrariness and imbalance. Web pundit David Weinberger has pointed out the cultural imperiousness of a classification system that devotes eighty-eight integers to Christianity, for example, while reducing the vast canon of Buddhist teachings to a single decimal point.[29] But Dewey was simply trying to develop a system capable of describing the body of published work available in North America at the time—a world in which Christian books vastly outnumbered Buddhist ones. And he freely admitted to valuing practicality above philosophical perfection. "Theoretical harmony and exactness has [*sic*] been repeatedly sacrificed to the practical requirements of the library," he wrote.[30] Even so, the system's fundamental rigidity has prevented it from evolving over time, making it seem increasingly like a relic of another era, what Weinberger has called "a Victorian sitting room with furniture that's too heavy to lift."[31] That criticism would likely have come as no surprise to Dewey, who recognized the philosophical limitations of any attempt at proposing a universal classification scheme. "Theoretically, the division of every subject into just nine heads is absurd," he wrote. "Practically, it is desirable."[32]

The Dewey Decimal System soon became the de facto standard in public libraries across the United States, thanks to Dewey's relentless efforts and the ongoing willingness of library administrators to submit to standardization and mechanistic efficiency in the service of progress. This approach also played nicely into a rising democratic ideal. Whereas once books and libraries had been the province of the educated middle and upper classes, public libraries were emerging everywhere (thanks in no small part to the philanthropy of Andrew Carnegie) to serve the needs of the common educated reader. Public libraries strove to serve the ordinary citizen rather than the elite scholar. That proletarian ideal often influenced the structure of the card catalog itself, which became a fixture of public life in the United States and, eventually, in Europe. And soon enough, card catalogs started to spread beyond the confines of the library world.

As the Industrial Revolution unfolded during the latter half of the nineteenth century, organizations of all stripes started to recognize a pressing need to control their intellectual capital. Many of them began to see the potential value in the card catalog as a "universal paper machine" (to borrow scholar Markus Krajewski's phrase).[33] Over time, corporations, government agencies, foundations, and other institutions began to adapt the card catalog for their own uses. It provided a stable mechanism for storing information, interpolating one piece of data against another, and transferring records from one location to another. Organizations began to discover the operational benefits of filing, indexing, cross-referencing, and making data available for immediate retrieval. And, like so many other information technologies, the card catalog delivers its real value at scale. As Krajewski writes, "As soon as a box of index cards reaches a critical mass of entries and cross-references, it offers the basis for a special form of communication, a proper poetological procedure of knowledge production that leads users to unexpected results."[34] Those results—what today we might call business intelligence—offered the prospect of a powerful competitive advantage for corporations, which in turn began to see the long-term potential in applying the lessons of the library catalog to their business operations. They found a willing partner in Melvil Dewey.

In 1876, Dewey established a company called the Library Bureau, whose mission was to sell to libraries and other organizations supplies such as catalog cards, drawers, "bureau boxes," and other material to help companies implement his scheme. The company also encouraged businesses to adopt its index-card systems for their internal record keeping, claiming that they would realize efficiency gains of up to 50 percent. By 1893, the firm had opened a new office in Chicago and had plans to expand to London and Philadelphia. In 1895, the Library Bureau invested in its own factory, expanding from the production of index cards into a wider range of business equipment designed to help organizations of all stripes manage their collective intellectual capital: cabinets, trays, embossing equipment, and a range of other office supplies. As government bureaus and corporate offices found themselves increasingly awash in paper—spurred by the adoption of new technologies like stenography, carbon paper, and typewriters—the Library Bureau saw the opportunity to insinuate itself into the organizational mainstream, recognizing that the innovations it had brought to the library world could have application across a broad swath of business and governmental enterprises.

In 1896, the firm formed a partnership with Herman Hollerith's Tabulating Machine Company (TMC), the company that several years later would change its name to International Business Machines, or IBM. Dewey signed a three-year contract with TMC to develop technologies to support government census operations, railroads, insurance companies, and other large-scale enterprises. The Library Bureau would provide the cards, and TMC would provide the equipment to process them. By this time Dewey had extricated himself from day-to-day management of the company but retained a significant ownership stake and the ceremonial title of president (though he drew no salary). He also continued to exert a powerful hand on the company's high-level strategy, insisting that the firm align its efforts with his interests in library, metric, and spelling reform. Thus Dewey played a direct role in establishing one of America's corporate pillars. And the card catalog, whose origins stretched back to Conrad Gessner's experiments with paper slips, thus insinuated itself into the organizational make-up of one of the technology giants of the twentieth century—the same company that, decades later, would invent the technology that would make card catalogs obsolete: the relational database.

By the end of the nineteenth century, the second Industrial Revolution had transformed the intellectual landscape of Europe and North America. The old paper tools—books, journals, and hidebound libraries—no longer seemed up to the task of managing the outpouring of published knowledge. Too much data were being produced for any one institution—let alone an individual—to manage, even with a card catalog. As the pace of innovation continued to accelerate, it brought social, economic, and cultural transformations that would thrust these newly industrialized people into a period of wrenching upheaval: like the people of Babel, scattered across the earth without a common tongue.

Into this changing world, on August 23, 1868, Édouard and Maria Otlet welcomed their son Paul.

2

The Dream of the Labyrinth

Some nations are born out of kinship, others out of conquest. The nation of Belgium was born from a song. In 1830, a few hundred spectators gathered in Brussels to take in what was by all accounts an unusually stirring performance of the French grand opera *La Muette de Portici* (*The Mute Girl of Portici*). The opera's patriotic message—with its rousing song "Sacred Love of the Fatherland" ("Amour sacré de la patrie")—struck a chord with the French speakers in the crowd, who like many of their compatriots had long felt aggrieved under generations of Dutch rule. The performance lit a fuse in the audience, igniting a nationalist fervor that swept through Brussels as the attendees streamed out of the theater and into the streets, where thousands soon joined in a massive protest for independence. The riots spread across the country, and within a year the fledgling nation of Belgium had declared its independence from the Netherlands.

As Europe's newest nation-state, Belgium seemed imbued from the start with a spirit of forward-looking idealism. After forming a national congress, the government invited a German prince named Leopold Georg Christian Friedrich to ascend the throne of its newly forged constitutional monarchy. At first, the fledgling kingdom struggled to find its footing. After severing ties with the Netherlands, Belgium lost access to the world markets commanded by the massive

Dutch fleet. As a result, the new regime faced considerable ill will from its powerful mercantile class. By 1845, fully one-third of the population of Flanders was living off charity.[1] But the country had considerable assets at its disposal: raw materials in the form of enormous coal and iron deposits in the south and east of the country, as well as a major port in Antwerp. The country was ripe for the technological revolution that would soon transform it into the most industrialized nation on the continent.

Eager to prove his bona fides as a leader, King Leopold oversaw the construction of continental Europe's first railway. Between 1840 and 1880 the railway network would expand by a factor of ten, faster even than the British railways.[2] Meanwhile, new industrial mining techniques allowed Belgian firms to tap the country's mineral resources, which found ready markets in the expanding industrial economies of France and Germany. The booming Belgian business climate of the mid-nineteenth century spurred a generation of ambitious young men to seek out new opportunities. Among their ranks was Édouard Otlet, a businessman's son who grew up in the aftermath of Belgian independence. His son Paul later described him this way: "My father was part of the generation that came of age immediately after the 1830 revolution," he wrote. "He had a clear vision—guided by principles of economic policy—that expanding the country's frontiers could improve its prospects."[3] Édouard Otlet spent most of his twenties doing exactly that, establishing a company that sold tram systems to cities around the world, in the process traveling as far as Madrid, Odessa, Alexandria, and Algeria. He amassed a small fortune, acquiring several chateaus and villas in Belgium and France and building the credentials and connections that would one day enable him to join the Belgian Senate.

In 1866, Édouard Otlet married Maria Van Mons, the daughter of a prominent Brussels family. Soon enough the couple started a family and seemed destined for a happy and prosperous life. That happiness was not to last. Maria died in childbirth in 1871, at the age of twenty-four, when young Paul was only three years old. A widower at twenty-nine, Édouard pressed ahead with his business, traveling the world while servants and tutors raised his two sons, Paul and Maurice. He refused to enroll his sons in school, convinced that a classroom education would hinder their intellectual development. Instead, he toured his family extensively around Europe and bought a share of an island in the Mediterranean, the Île du Levant, where Paul continued his private studies: playing piano, learning to dance, and holing up with his books between excursions on the family yacht. Throughout his travels, Paul usually kept a diary and often collected assorted rocks, plants, and other found objects, which he ultimately put on display in his so-called Musée d'Otlet, which occupied part of the first floor of the family home in Brussels—the beginning of his lifelong fascination with museums. In a further glimpse of things to come, he and his brother drew up detailed plans for a so-called Limited Company for Useful Knowledge.[4]

Ensconced in this privileged life, with few playmates other than his younger brother, Paul developed into a painfully shy and preternaturally serious young man. From an early age, he took refuge in the family library, preferring the company of books to people. That solitary disposition would persist even after he finally started school at age fourteen, when his father enrolled him at the Jesuit Collège Saint Michel in Brussels, where the socially maladroit Paul endured constant mockery from his schoolmates, further deepening his isolation.

The diary that he kept obsessively from the age of eleven to twenty-seven reveals an intense young man struggling to find his purpose in the world. "My mind is continually tormented," Paul wrote at the age of seventeen. "I don't know where to turn next with my career. Such is the problem that is ceaselessly buzzing in my ears. Indecision is the worst of evils."[5] At school he continued to immerse himself in books, writing detailed notes about his classes and indulging in typically grandiose adolescent reflections about the nature of the world. He made few close friends. "I don't like society, above all this frivolous and shallow society," he wrote. "Literature suits me.... Reflection and meditation are the things I hold dear."[6]

One October day in 1885, one of the school's Jesuit fathers took Otlet aside and offered him an opportunity that he thought might well suit the introverted young man. "Today during study hour, he called me up and asked me to be the librarian," Paul wrote. "I accepted with pleasure."[7] Twice a week, he would handle students' requests for books, using the rest of the time to read whatever he chose and spend time perusing the library catalog. "It seemed a

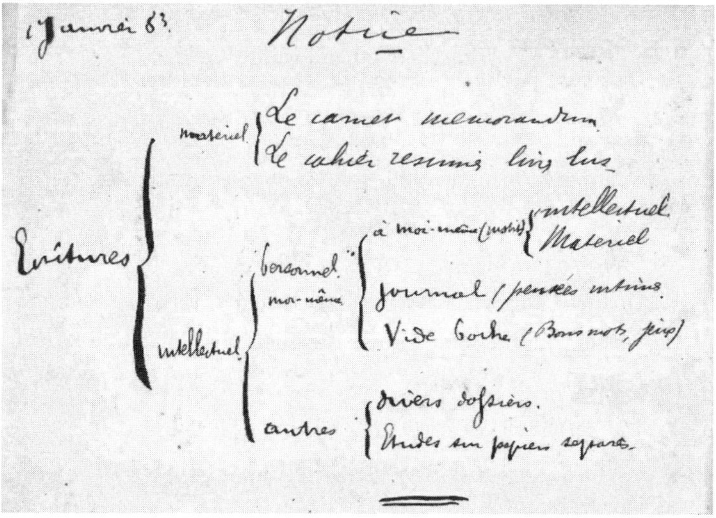

Otlet's personal library classification scheme, developed in 1883 at age fifteen. Reproduced with permission from the Mundaneum.

wonder," he later wrote of the library catalog, "this instrument that allowed me to use all of these books."[8]

The Jesuit fathers could scarcely have chosen a more ideal candidate to take over the library. By the time he started working in the school library Otlet had already started to develop an elaborate classification scheme for his own personal collection, which he outlined in a diary entry from 1883:

Material
- –Memoranda, notebooks
- –Summaries of books read

Intellectual
- –Personal
  - –For myself (intellectual/material)
  - –Dairy (intimate thoughts)
  - –Pocket books (witty sayings, amusing ideas)
- –Others
  - –Various topics
  - –Studies of other documents

Otlet also paid attention to the physical organization of his collection, developing a scheme of file folders, drawers, and boxes to contain his collection:

1. Dossier—to place everything that should be classified
2. Papers with the same format, different things placed in files (cartons)
3. Boxes of things (souvenirs)
4. Drawer for literature (by others)
5. Drawer for me (personal)[9]

If the classification seems a bit anachronistic—Borges's famous satirical classification of the animals comes to mind, including "those that, from a distance, look like flies"[10] (see chapter 9)—nonetheless it reveals a youthful determination to fit a chaotic world into an orderly chain of concepts. Writing in his diary at age fifteen, Otlet reflected on the impetus behind his burgeoning urge to classify his intellectual world. "My God! What a feather brain I am, always on to something new, beginning and never finishing anything. I write down everything that goes through my mind, but none of it has a sequel. At the moment there is only one thing I must do; that is, to gather together my material of all kinds, and connect it with everything else I have done up till now."[11]

Otlet took notes on whatever he was reading, jotting down passages that caught his attention and—like Conrad Gessner—storing them on individual pieces of paper that he then sorted into folders. "To gain time," he wrote, "instead of immediately developing a thought that one has read, one simply makes a note of it on a piece of paper which is put in a folder. On Saturday, for example, these papers can be taken up one by one for classification."[12]

In the years ahead, Otlet's classification would continue to evolve with his personal interests, eventually expanding to include such categories as Literature, Philosophy, Social Sciences, Law, Politics, and Moral Questionings. And like many adolescent boys, he also took a keen interest in news of the latest technological innovations. At the age of seventeen, he wrote excitedly in his diary about a new device that had captured his imagination: the telephone. "What a happy idea crossed my mind when I thought about using the telephone," he wrote. "It is an admirable invention."[13]

As his interest in science and technology deepened, Otlet also began to search for an inner purpose that he believed he could find only by moving closer to God. These twin impulses—toward scientific mastery on the one hand, and religious awakening on the other—converged and intertwined in Paul's developing psyche. Throughout his life, he would exhibit a remarkable capacity for painstakingly detailed analytical work, coupled with a deeply held belief in spiritual transcendence.

As he continued his bibliographical journeys, Otlet eventually butted up against the limitations of the school's library collection, confined as it was to books the Jesuit fathers deemed appropriate. Seeking to broaden his intellectual horizons, Otlet—like his Renaissance forebears—gravitated toward dictionaries and encyclopedias. He especially liked Pierre Larousse's *Grand dictionnaire universel du XIXieme siècle.* Published from 1866 to 1876 and spanning fifteen volumes with more than 20,000 total pages (with another two supplements published after Larousse's death in 1876), it offered an endless source of diversion for Otlet as he whiled away his time in the school library. "These were my best friends," he wrote. "They made it possible to be curious about anything, without having to track down a book on the subject."[14] At length he began to experiment with creating his own personal encyclopedia, a grand adolescent scheme that, though abandoned, planted the seed for the far grander projects to follow.

Growing up in this cocoon of private schools and foreign holidays, with his nose usually buried in a book, Otlet may have had few inklings of the tumultuous changes that were starting to reshape the outside world.

On February 26, 1885, representatives from fourteen European nations concluded a marathon three-month-long conference in Berlin, during which they attempted to sort out their competing claims in Africa. At the time, 80 percent of Africa remained unclaimed by the colonial powers, but the so-called Dark Continent had started to attract considerable interest all over Europe, thanks in part to the sensational reports of explorers like Henry Morton Stanley (of "Dr. Livingston, I presume?" fame). For some Europeans, Africa seemed like a fantastical world, a primordial wilderness promising mystery and adventure. For others, it looked like one big gold mine, both literally and figuratively. For the increasingly industrialized European powers, the temptations would prove irresistible.

After the Congress of Berlin, Great Britain came away with a string of territories stretching in an almost unbroken chain from the Cape to Cairo (including present-day Egypt, Sudan, Uganda, Kenya, Zambia, Zimbabwe, and South Africa, as well as Nigeria and Ghana); Germany staked its claim in East Africa (Namibia and Tanzania); while France consolidated control over much of western Africa (Chad, Mauritania, and French Equatorial Africa). Up until then, the European powers had confined themselves largely to the African coast. But now they hoped to penetrate the interior, superimposing new territorial boundaries that bore little relation to the existing cultural boundaries under which the indigenous African population had lived for generations. The great European Scramble for Africa, as it has come to be known, would set in motion a series of geopolitical conflicts that would reverberate for decades to come—not just in Africa, but all over the world.

Belgium also took part in the Congress of Berlin. King Leopold's son, King Leopold II, had ascended the throne in 1865, bringing with him a zeal for colonial expansion. "Belgium doesn't exploit the world," he once said to an adviser. "It's a taste we have got to make her learn."[15] King Leopold I had explored various colonial opportunities—at one point even trying to annex part of Texas—but those efforts ultimately came to naught. Leopold II took up the cause with gusto and, soon after he took power, began casting about for potential candidates. After flirting with the prospect of claiming colonies in the Philippines, Argentina, or Fiji, the young king plunged Belgium headlong into a colonial adventure in Africa. Those ambitions would ultimately lead his nation into one of history's great imperial misadventures: the Congo.

Soon after securing control over the Congo in Berlin, Leopold II announced the formation of the Congo Free State, a new administrative entity entirely under his control. Leopold saw the enormous economic potential of securing access to the Congo's rich lode of raw materials. In the early years of the colony, Belgian traders brought back a steady supply of ivory, used for ornamental carvings, hair combs, and keys for the factory-made pianos that were fast becoming a fixture in middle-class American and European households. By the late 1890s, however, the ivory market would be eclipsed by a booming worldwide demand for rubber. A rapidly expanding global network of telegraphs, telephones, and electrical lines were fueling demand for rubber insulation to protect the wires; factories were also discovering uses for treated rubber in hoses, tubing, and gaskets. Soon, automobiles would arrive, and with them an even greater demand for rubber. The Congo's vast rain forest, containing plentiful rubber vines, blanketed fully half of the colony.[16] And an indigenous population of loosely organized tribes provided a ready source of cheap labor, which the Belgian Force Publique was quick to enlist in the effort through conscription and even more brutal forms of coercion.

Leopold II cloaked his economic motives in Kiplingesque rhetoric about the white man's burden by vowing to improve the lot of the natives and using his formidable rhetorical skills to persuade his loyal subjects of the nobility of the project. That spirit of noblesse oblige held sway across much of Europe, where such high-minded sentiments obscured the rapacious looting and mistreatment of indigenous populations from public view. From his royal soapbox, the king promoted a program not of economic conquest but of charitable and scientific outreach. "To pierce the darkness which hangs over entire peoples, is, I dare say, a crusade worthy of this century of progress,"[17] he said at the first meeting of the International African Association, an organization formed in the wake of a conference on Belgium convened by Leopold, whose chairmanship the king had then graciously agreed to accept.

Spurred on by the king's rhetoric, Belgium launched its first civilian expedition to the Congo in 1886, organized and financed by none other than Édouard Otlet. The trip's ostensible goal was to gather indigenous artworks for a planned museum; the results proved disappointing. The trip's leader did return with the son of a Congo chief, a young man called Mayalé, who went on to work as a servant in the Otlet household. Otlet's later expeditions were overtly commercial in nature and came in the wake of a series of secret government-sponsored expeditions financed by Leopold and led by the world-famous Henry Stanley. Officially, the king framed these expeditions as purely exploratory in nature. But in his private correspondence he freely acknowledged his ulterior motives. "I'm sure if I quite openly charged Stanley with the task of taking possession in my name of some part of Africa, the English will stop me," he wrote. "So I think I'll just give Stanley some job of exploration which would offend no one, and will give us the bases and headquarters which we can take over later on."[18] Soon enough, Stanley had helped the king engineer the acquisition of the Congo.

Almost from the start, the Free State of Congo was a star-crossed venture, one that would provide a cautionary example for the other European powers locked in a race to expand their imperial reach. After winning control of the Congo, the king established the Rubber Trading Colony, which in the decade that followed would become the cause of massive brutality and exploitation. Some estimates have suggested that as much as half the population—an estimated 10 million people—died as a result of the Belgian occupation.[19] The semiprivate imperial venture, which had initially nearly plunged Leopold into bankruptcy, would start to throw off wildly lucrative profits. Fueled by these riches, Belgium would propel itself into a central role in European affairs.

It took years for the truth of what happened in the Congo to come to light, and when it did it was thanks to the work of pioneering crusaders, such as Edward Morel, and prominent writers like Sir Arthur Conan Doyle and Mark Twain, who joined the cause with righteous enthusiasm after learning of the atrocities. In 1905 Twain published a pamphlet entitled *King Leopold's Soliloquy*, a satirical apologia in which a fictional Leopold II tries to justify his rapacious ways, railing against American missionaries, British consuls, and other "tiresome parrots" who turned a harsh light on the Congolese atrocities. Twain delivered a withering assessment of what he saw as craven profiteering on the king's part. "It seems strange to see a king destroying a nation and laying waste a country for mere sordid money's sake. . . . We see this awful king, this pitiless and blood-drenched king, this money-crazy king towering toward the sky in a world-solitude of sordid crime."[20]

History has rendered a harsh judgment on Leopold II for the humanitarian disaster of the Congo, which in hindsight overshadowed his many other achievements. His reign saw a great flourishing of the arts—notably the rise of avant-garde literature and theater—as well as steady progress in science and technology. Along with Brussels mayor Jules Anspach, he initiated an aggressive building campaign in hopes of transforming the city into a world capital.[21] While these accomplishments may have endeared him to many of his subjects at the time, his misdeeds in the Congo have largely defined his historical legacy outside of Belgium. In *King Leopold's Ghost*, Adam Hochschild describes him as "a man as filled with greed and cunning, duplicity and charm, as any of the more complex villains of Shakespeare."[22] Historian Barbara Emerson paints a somewhat more sympathetic picture of the Belgian king, depicting him as a controlling and calculating figure, but far from the monster that many of his critics have made him out to be. On first hearing reports of atrocities in the Congo, in fact, he appears to have reacted with disgust. "These horrors must end or I will retire from the Congo," he reportedly said. "I will not be spattered with blood and mud."[23] But while he had originally pursued the Congo for political reasons, he now seemed to have given in to the temptations of the vast wealth that had started to flow into his personal coffwers. Consumed with greed, he ultimately succumbed to what Emerson calls "a frightening case of moral decadence."[24] When the foreign press took up the issue as a cause célèbre, Leopold II saw the campaign as a British-led assault on his power, one that required a strong and unequivocal response. Rather than address the humanitarian concerns, however, he chose to adopt a defensive posture, denying the allegations and insisting on the nobility of the Congo project.

The Belgian Congo might seem far removed from the quiet life of an adolescent Paul Otlet puzzling out his schemes for the library catalog. But directly and indirectly the Congo project would influence Otlet's life and work for years to come. Perhaps because of his father's involvement, Paul took a particular interest in the Congo. In 1888, twenty years old and still enrolled at the University of Leuven, he published a pamphlet entitled *Africa for the Blacks* (*L'Afrique aux noirs*), in which he gave voice to the same spirit of civilizing the natives that had characterized King Leopold II's early proclamations. Like many of his countrymen, Otlet viewed the king as a great visionary, though he opposed the colonialist policy. Instead, he advocated repatriating American slaves in Africa—an idea that, while it might seem odious from a contemporary vantage, held sway throughout the nineteenth century; a number of prominent Americans had embraced the notion over the years, including James Madison, Thomas Jefferson, James Monroe, and Abraham Lincoln. By the end of the nineteenth century the movement had gained plenty of influential adherents among the political and business classes. Otlet argued that repatriated slaves offered the best hope of transplanting the values of civilization onto the African continent, proposing that "the vast Independent State of the Congo would open its doors to these fellow American citizens who are their children."[25] He presumably had little idea of the fate that would likely befall those former slaves were they actually to go to the Congo; like his king, he never set foot there. Nonetheless, Otlet would remain deeply engaged by the "African question" for years to come, urging European nations to help improve the lot of the natives. "The Congo project is primarily a Christian and humanitarian undertaking," he wrote in his pamphlet, arguing in white-man's-burden style that Africans were "men and brothers who we must relieve of their long moral and intellectual decay." While he seemed to participate fully in the imperial spirit of noblesse oblige that animated so many imperial projects, Otlet also seemed to recognize the limits and the perils of colonial projects. "The history of all social movements shows us that one must be careful of rapid progress without adequate transition, and that all settlements thus established must be infused with native blood."[26] That sentiment would prove all too prophetic.

Long after the atrocities in the Congo had come to light, Otlet continued to see King Leopold II as a visionary. In 1927, he penned a tribute to the late king (who had died in 1909), in which he acknowledged the problematic aspects of the Congo project but nonetheless judged the king a master "sociologist" and celebrated him as "a great man whose memory we must keep." Leopold was "a king of big ideas and grand visions . . . a Worker, a Builder, a Man of Accomplishment."[27] Perhaps Otlet's fealty stemmed from an instinctive nationalist loyalty, or perhaps it had something to do with the king's later support for some of his key projects. Otlet may also have seen in Leopold II a kindred spirit: a lonely man striving to realize an expansive vision in a sometimes hostile world.

Otlet and Leopold II shared a conviction in the superiority of European culture—and in this they were scarcely alone. Throughout the late nineteenth century, many Europeans grew up with the received wisdom that their culture represented the pinnacle of human progress. Western technological and scientific advances were improving the lot of people the world over; the path to a more prosperous future involved exporting a set of underlying cultural values—not to mention people, products, and technology—abroad. The march of progress signaled the continued advancement of human knowledge toward the universal. Indeed, the possibility of universal order almost amounted to an article of faith. It was, as Boyd Rayward writes, a time of great "cultural certitude."[28]

Otlet's early attempts at both library cataloging and polemics reflect the assumption of cultural superiority that was endemic in Europe during that time. Europeans saw themselves as inhabiting an "advanced" civilization, in marked contrast to the "primitive" civilizations of Africa and other colonial settlements. Notions of cultural relativism remained a long way off in the late nineteenth century. The ideal of unidirectional progress undergirded much of the universalist and internationalist rhetoric that began to take shape during this era. The prospect of reforming the African nations went hand in hand with ideas about cataloging books: Both impulses reflect what his biographer Françoise Levie calls "the Otletian will to reorganize the world."[29]

Otlet's convictions about liberating the African people through an enlightened form of self-government stemmed in part from his proclivity for positivism, a worldview that he first encountered in his youth and one that would shape his life and work. Rooted in the writings of Auguste Comte, positivism argued for a strictly scientific view of the world, asserting that the only valid truths are those that can be verified through empirical observation. Just as the physical world operates under a set of inviolable laws—like gravity—human society also operates under a set of laws. Positivism asserted that all cultures move through progressive stages of development: first theological, then metaphysical, and finally "positive." In the first stage, according to Comte, God (and, by extension, the church) reigned supreme over humanity; divine edicts and articles of faith trumped individual agency. In the so-called metaphysical stage, societies tried to correct that balance by instituting concern for the individual and embracing a belief in inalienable human rights. Nonetheless, the power of the state safeguarded those rights on behalf of the people. In the third stage—yet to come—societies will progress to the scientific, or positive, level of development. Human knowledge will be set free from its old institutional constraints, and people will at last begin to govern themselves.

In addition to his theory of social evolution, Comte also proposed what he called an "encyclopedic law" that governed the classification of human knowledge, in much the same way that the material world and human society followed a set of laws. All human knowledge, in his view, fell into one of five clearly demarcated categories: astronomy, physics, chemistry, biology, and sociology. This last category, "the true science of Man," constituted the pinnacle of human understanding, "the last gradation in the Grand Hierarchy of Abstract Science,"[30] promising to unify the pursuit of abstract knowledge with the forward march of civilization. Thus, for Comte, the pursuit of a universal classification scheme went hand in hand with a belief in the possibility—indeed, the inevitability—of social change.

If one accepts—as Otlet did—the positivist premise that cultures evolve in successive stages from primitive to advanced, it stood to reason that more advanced societies could help lesser societies move forward along the path to realization. Indeed, Comte's work deeply influenced Karl Marx, who believed in social evolution toward a stateless society. And it inspired Otlet to devote much of his life to pursuing the classification of human knowledge as an integral component of a much broader utopian project.

In 1889, Otlet wrote down a personal credo. "I believe in the great principles of positivism and evolution," he wrote, "the relativism of knowledge and the historical formation of concepts."[31] Utopian dreams are often an occupational hazard for librarians. Otlet saw little distinction between creating a new classification of human knowledge and reorienting the world's political system. Just as human civilization could advance from primitive tribal societies to sophisticated nation-states, and knowledge progress from oral culture to a kind of encyclopedism, so too the countries of the world could one day advance toward a postnational, global order, by embracing what Otlet called "international life."[32]

After graduating from the Free University of Brussels—where he had transferred after leaving the University of Leuven and spending six months intellectually soul-searching in Paris—Otlet had to turn his attention away from diary-keeping and essay-writing and toward the more pressing business of finding a way to support himself and his young family. In 1891, he had married his cousin Fernande, whom he had courted assiduously for several years after first encountering her at a family gathering. Fernande finally acquiesced to his proposal, and soon enough the couple would welcome their first son, Marcel, named after Paul's little brother. Now with a family of his own, Otlet seems to have recognized that he would need to sacrifice his idealism for the sake of a career. His father's fortunes had taken a turn for the worse thanks to a series of bad investments, leaving his substantial

Paul Otlet as a young man (c. 1890). Reproduced with permission from the Mundaneum.

subsidy of 12,000 francs a year in question. "Now all our fortune is gone, perhaps momentarily, forever perhaps. . . . It is necessary to work."[33]

Otlet took his father's advice to go to law school, hoping to secure safe passage to the comforts of a bourgeois life. If he applied himself, perhaps he might one day have hoped to follow his father's footsteps into the Belgian Senate. With his father's support, he joined the law practice of Edmond Picard, an influential Brussels attorney and old family friend who was also the first socialist member of the Belgian Parliament. Picard soon put Otlet to work in helping assemble a massive compilation of Belgian law dubbed the *Pandectes Belges*. The young lawyer hated the work. Whereas he had at first hoped that practicing law might give him a way to put his positivist ideals into practice in the political arena, he discovered that much of his work involved, as he put it, "stupid acts of procedure." Within a year, Otlet had decided to abandon the legal profession. "The Bar," he wrote in his diary, "what misery!"[34] With a household to support and no clear idea of what to do next, Paul returned to his first love and first consolation: the world of books.

3

Belle Epoque

At the Paris World's Fair—the Exposition Universelle—of 1900, nearly 50 million spectators crowded through the massive arched gateway of the Place de la Concord to marvel at the largest such exhibition ever staged, dwarfing the more famous Exposition of 1889 and the Chicago World's Fair of 1893. The fairgrounds sprawled across 360 acres in the heart of central Paris, with flags and pennants flying from almost every window as a steady stream of humanity flowed by the panoply of exhibits on display. The Exposition gave many attendees their first glimpse of such Industrial Age innovations as escalators, diesel engines, magnetic audio recorders, and Campbell's soup cans. Some visitors took in the attractions by riding a two-mile-long moving sidewalk that circled the fairgrounds, on which a few of them might have spotted a solitary figure sitting behind a large wooden camera, turning a metal crank: Thomas Edison.

Moving pictures played a prominent role in the Expo. In addition to Edison's recently invented Kinetoscope—a precursor to the modern movie camera—visitors could take in the first-ever public exhibition of talking pictures, thanks to an invention called the Phono-Cinéma-Théâtre; step inside a panoramic 360-degree motion-picture projector designed by the Lumière brothers; or attend a lecture by a Russian scientist named Constantin Perskyi, in which he introduced a word he had recently coined: "television." The motion-picture projectors, moving sidewalks, and ubiquitous electric lamps all drew their power from the Palace of Electricity, a massive pavilion that did double duty as both a public exhibit hall and the Expo's central power plant. The exterior of the building shimmered with millions of bulbs that were switched on at night, bathing the surrounding area in, as one guidebook put it, "a flood of fairy light."[1]

In another corner of the fair stood an amusement park–like attraction devoted to Old Paris, designed by the proto–science fiction writer Albert Robida. In 1883, Robida had written his first popular novel, a lighthearted futuristic romp called *Twentieth Century* (*Le Vingtième siècle*). The book depicts a technologically advanced world of underground trains, flying machines, moon colonies, and a vast underwater settlement devoted to gambling and other forms of entertainment: a kind of subaquatic Sin City. The storyline hinges on an imaginary device dubbed the téléphonoscope, an invention capable of transmitting words and pictures across great distances and rendering them on a screen, thus allowing people the world over to consume news and entertainment, as well as shop for merchandise from the comfort of their homes. Although the book reads as a playful flight of scientific fancy, it also points to the pervasive sense of technological optimism that captivated many European minds in the late nineteenth century. Along with a handful of other contemporary writers like Jules Verne and H. G. Wells, Robida mined the technological landscape to portray an optimistic new world whose inhabitants solved major social problems—like public education, communication, and international cooperation—through the skillful application of scientific discoveries. Robida's infatuation with technology would not last, however. His last book, *The Engineer von Satanas*, published in the wake of World War I, amounted to a withering rebuke of technological progress.[2] But in those hopeful years of the Belle Epoque, before Europe witnessed with horror the dark side of industrial innovation, Robida was scarcely alone in embracing the potential benefactions of science and technology.

That optimism was on full display at the Paris Expo. By 1900, Paris had grown into a truly modern city, one of just eleven in the world with more than a million inhabitants.[3] Its boosters hoped to transform it into a new Mecca of the Industrial Age, a living testament to technological progress. Electric lines fanned out across the city to government buildings, commercial establishments, and thousands of ordinary Parisian households—helping Paris live up

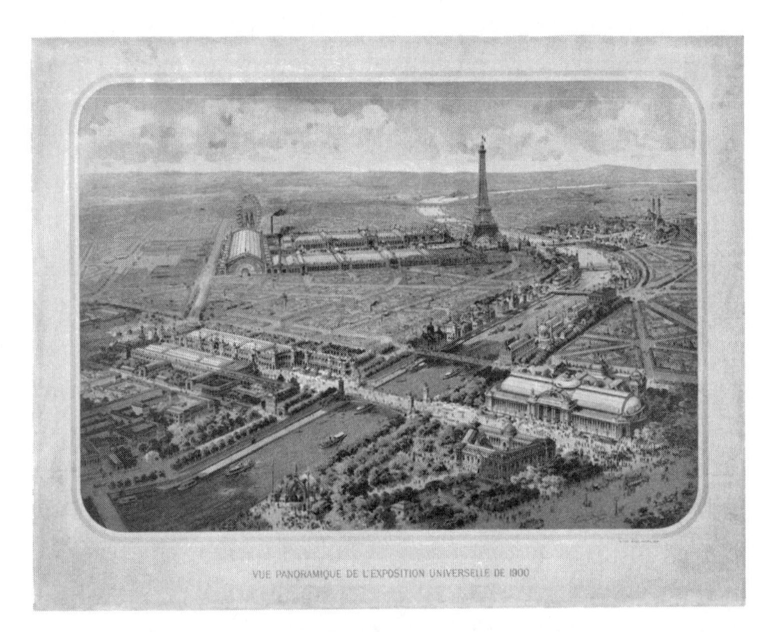

Panoramic view of the Universal Exposition of 1900 (*Vue panoramique de l'exposition universelle de 1900*). Baylac, Lucien. Library of Congress Prints and Photographs Division.

to its reputation as the City of Lights. Under Baron Haussman, the city had modernized its formerly medieval street plan, while below ground a network of pneumatic tubes snaked along for more than 120 miles, delivering millions of paper messages throughout the city on any given day.[4] Paris was becoming, as historian Rosalind Williams put it, a "unified system."[5]

That system radiated outward as well as inward, as Paris cemented its position as a cultural and economic hub. Trams and railway lines were cropping up everywhere, while an interlocking network of post offices, telegraph stations, and telephone lines was shrinking the time and distance required to conduct business across national borders. As a result, capital was flowing through the industrial world more fluidly than ever. Many organizations took advantage of this rapidly evolving infrastructure to secure access to new markets, growing larger and increasingly multinational. People were moving across borders as well, with tens of millions of Europeans crossing the ocean to find new work in America. Goods zipped back and forth from the New World thanks to faster ships, while improved roadways and motorized transportation were making it easier than ever for goods to circulate worldwide. Along with New York and London, Paris stood at the intersection of these emerging trends, rapidly transforming itself into an international urban center of a wholly new order.

Just as money, products, and people were moving fluidly across borders, so too was the currency of ideas. Printing technologies, coupled with an increasingly efficient postal system, allowed the literate classes to communicate with each other more easily than ever before. Scholars, long accustomed to working in their cloistered campuses, began to forge international associations that, in turn, bred new forms of communication: conferences, journals, and a growing market for scholarly monographs. Social and political movements, once largely constrained by geographical limits, began to take on international dimensions: pacifists, communists, feminists, theosophists, and believers in animal rights began seeking out likeminded souls elsewhere in the world. The closing years of the nineteenth century saw the rapid growth of international associations, as people with similar interests began to connect with each other through the newly globalized postal network. In the early 1870s, there were twenty-five international associations in the world; by the first years of the twentieth century, their number had risen to more than 600.

Scientists were the first to organize, but they were soon followed by other professionals working in the increasingly specialized fields that were taking shape in the wake of industrialization: for example, architects, engineers, bankers, and hoteliers.[6] By one estimate, the number of available professions quadrupled from 1880 to 1911.[7] As these professions started to coalesce into organizations, they started to gather for international conferences (or "congresses," as they were known). This growing enthusiasm for conference going reflected the hopeful spirit of the times. "International congresses are the outcome of modern civilization," as one geographer had put it a few years earlier.[8]

In addition to serving as a venue for knowledge sharing and collaboration, the nineteenth-century congresses also played an important role in helping to establish international standards for a wide range of pursuits. Scientists, for example, needed a commonly agreed set of weights and measures; doctors needed a common vocabulary for medical conditions; bankers needed a common language to differentiate their financial instruments; printers needed to agree on standard paper sizes, and so on. International associations seemed naturally suited to play such a role. But this new level of transnational cooperation signaled more than just a practical alliance; underlying these associations was a spirit of universal human endeavor—with strong echoes of Comte's positivism—that "spoke to an idealized vision of science and technical knowledge," writes historian Mark Mazower, "a creed without borders."[9]

In 1900, delegates from all over the world congregated in Paris to take part in a series of august-sounding international congresses dedicated to a bewildering array of topics: psychology, aeronautics, meteorology, vegetarianism, homeopathy, alpinism (focused on the protection of mountain plants), beverage-yielding fruits, numismatics, naval architecture, arboriculture, and pomology, to name just a few.[10] Whereas the subsequent U.S. world's fairs of the twentieth century would hinge largely on corporate-sponsored pavilions that served largely as promotional endeavors, for many attendees the Paris Expo evoked the atmosphere of a great academic convocation, taking place amid the electrified spectacle of the Parisian cityscape.

Europe's increasingly interconnected scholarly and professional communities gave rise to a new transnational sensibility that came to be known as "internationalism," a term that had only come into existence a few decades earlier, coined by the British philosopher Jeremy Bentham. Now, a new internationalist movement, coupled with—and enabled by—new communications technologies and a spirit of cultural progress, imbued many learned Europeans with a new way of thinking about world affairs, one less rooted in traditional national concerns and more concerned with building a futuristic, postnationalist global society.

The same spirit of internationalism drove a surge of interest in another star of the 1900 Exposition: Esperanto, invented by a Russian ophthalmologist named Dr. Ludwig Zamenhof. The Expo marked a kind of coming-out party for the new "universal second language." Like Otlet, Zamenhof harbored utopian ideals, hoping that one day Esperanto would enable citizens of different countries to communicate more easily, and in so doing contribute to international peace. Years later, in *Mein Kampf*, Hitler would denounce Esperanto as a tool of the "International Jewish Conspiracy." The Nazis specifically targeted Zamenhof's family for extermination.[11] Paradoxically, the rise of utopian internationalism, which found such powerful expression at the Paris Expo, coincided with a surge of nationalist fervor—one that would ultimately drive the continent's population into two cataclysmic wars.

One might well look at the idealism of the Expo through the lens of Jane Jacobs's famous critique of modernist urban planning: as "a dishonest mask of pretended order."[12] Certainly the imagined world on display there presented a relentlessly rosy picture of progress, papering over the dark sides of innovation and "progress"—like industrial militarization, and the economic exploitation of less developed countries—that would soon come into focus during World War I. Nonetheless, the Expo offers a window onto a moment when optimism about the possibilities of international cooperation seemed triumphant.

At the opening ceremony, the French minister of commerce Paul Delombre gave a speech celebrating the wonders of the technological transformation on display. "Words fail me to express the grandeur and the extent of this economic revolution," he said. "We have seen the forces of nature subdued and disciplined." "Machinery," he added, "has become the queen of the world."[13] He also remarked on the role communications networks were playing in shrinking the distances between nations and fostering the possibility of collaboration. "The telephone, that sorcerer, brings to our ear the words and even the tone of a friend's voice, separated from us by hundreds of miles. The intensity and the power of life—of death itself—recoil before the victorious march of the human soul."[14] After the minister finished his speech, a brass band struck up a tune as a procession of dignitaries walked up a grand staircase lined with guards in white trousers, leaving, as one observer put it, "like a scene from a grand opera."[15]

A few steps away, in the Grand Pavilion, thirty-two-year-old Paul Otlet and his partner Henri La Fontaine had set up an exhibit to demonstrate their own project, a so-called Universal Bibliography (*Répertoire bibliographique universel*). With his pince-nez and wispy beard, Otlet may well have been trying to strike a professorial demeanor as he lectured visitors about his creation. For the previous eight years, he had been working tirelessly with La Fontaine to assemble a collection of more than 3 million index cards, cataloging a vast number of published books and other sources, all classified by author and subject. The exhibit included accompanying charts and graphs depicting their evolving vision of a global catalog that would provide access to the entirety of human knowledge.

Otlet first met La Fontaine during the former's short-lived legal career, when both men worked for Edmond Picard on his massive compilation of Belgian law. They seemed particularly well suited to the task, sharing as they did a passion for books and bibliography. They also soon discovered a mutual interest in Comte's positivist philosophy.[16] La Fontaine was Otlet's senior by fourteen years, yet the two men nonetheless bonded almost immediately. A tall, thin man with high cheekbones and pensive gray eyes, La Fontaine possessed an aristocratic demeanor. He had by all accounts a remarkably magnetic personality. A committed socialist, pacifist, and feminist—not to mention a Freemason—he had founded the Belgian League for the Rights of Women in 1890 and helped form a Society for Social Studies. When Otlet first met him, La Fontaine, thirty-eight, was a confirmed bachelor who still lived at home with his sister.

Henri La Fontaine. Reproduced with permission from the Mundaneum.

In 1893, the two men decided to form a new association, an offshoot of the Society for Social Studies that they dubbed the International Institute of Sociological Bibliography. Inspired by the Comtean dream of developing a "true science of Man"[17] as a force for social good, they began work on a bibliographical survey of sociological literature that they hoped would serve as a foundation for a grand synthesis of knowledge about society. In 1895, they broadened the scope of their project, extending their reach beyond the realm of sociology alone to begin cataloging published information on a wide range of topics. That year, they established a new organization to support their efforts, dubbed the International Institute of Bibliography (*Institut International de Bibliographie*, or IIB).

Conrad Gessner had created his great bibliography by cobbling together material from a wide range of existing sources. Otlet and La Fontaine followed his example, building their Universal Bibliography largely by drawing on previously published material. The men would spend hours rifling through booksellers' catalogs, published bibliographies, and other sources, cutting up each entry into slips of paper and pasting the contents onto an index card, which would then be assigned a classification number—based at first on the Dewey Decimal System—and stored in a drawer. In the early days, La Fontaine's sister, Léonie, helped them create entries for the catalog, which, by the end of 1895, numbered more than 400,000 entries. Eventually, Otlet and La Fontaine agreed that this method seemed too haphazard. So they decided to invest in an expensive new piece of equipment: a typewriter.[18]

Otlet's and La Fontaine's early attempts at launching their venture met with resistance and, occasionally, even ridicule. After the pair convened their first conference on international bibliography in 1895, they met with near-universal skepticism from other European librarians and bibliographers, who rose up in what Rayward

The Universal Bibliography (*Répertoire Bibliographique Universel*), c. 1895. Reproduced with permission from the Mundaneum.

describes as "elegant tumult" against the brash young Belgians: "Nationalist and professional fervor surged through their pens, as though from the threat of some aggressive imperialism."[19] Perhaps it should come as no surprise that the keepers of Europe's library catalogs were in no rush to embrace a pair of unknowns from the Low Countries asking them to cede their bibliographical sovereignty. Some objected to the project on its most basic premise, arguing that true universality across disciplines was an unachievable dream; better to have subject-specific schemes tailored to the nuances of each discipline rather than one shallow classification that tried to cover every subject under the sun. Others took issue with the project's embrace of the Dewey Decimal Classification, which they deemed too simplistic and culturally biased (relegating all of European history to a single digit, for example, thus putting it on a par with lesser continents like North and South America). The eminent French bibliographer Henri Stein castigated Otlet and La Fontaine's project as "grandiose and rather temeritous."[20] But the partners pressed ahead, traveling to conferences in England, France, and Italy to muster support for their project.

The newly founded institute gained much-needed credibility when La Fontaine won a seat in the Belgian Senate as a founding member of the recently formed Labor Party (a position he would hold for thirty-six years). La Fontaine used his status to promote the effort, prevailing upon the well-known Belgian chemist and industrialist Ernest Solvay to support its activities, and convincing the Belgian Ministry of the Interior to fund a related entity called the Office of Institutional Bibliography. The two intertwined and similarly named organizations—the Office of International Bibliography (OIB) and the International Institute of Bibliography (IIB)—would operate in tandem for years to come; Otlet frequently used the terms interchangeably.[21]

Later that year, the men submitted their proposal to King Leopold II, wrapping their appeal in rhetoric that they knew would play to the king's proclivity for projects of high purpose. "In taking this initiative," Interior Minister François Schollaert wrote to the king at Otlet and La Fontaine's behest, "you can establish in your country the principal organ of intellectual life."[22] For the imperially minded monarch, the prospect of serving as royal patron to a universal library would likely have held much the same appeal as it did to the emperors of Sumeria, China, and Alexandria. By royal decree, he endorsed the OIB on September 12, 1895.

Having thus acquired for their organization an air of governmental authority, La Fontaine also helped Otlet master the arts of politicking and persuasion. Through a combination of determination, guile, and relentless personal networking, the two men began to insinuate themselves into the center of a global dialogue about the future of books and libraries: convening congresses, issuing bulletins, and promoting their project at every available opportunity. La Fontaine described the IIB's mission as "nothing less than a question of creating a world depot, where all human ideas can be automatically stored in order to be spread afterwards among people with a minimum of effort and a maximum of effectiveness," going on to describe it as "a central institute where all those who hope to collaborate in the progress of humanity, will be able, immediately and mechanically, to obtain the most detailed and complete information."[23]

With their new royal imprimatur, Otlet and La Fontaine began to hire full-time staff for the project, securing offices in the Palais des Musées Royaux in Brussels, where they opened their growing index card catalog to the public every day during business hours. They also established a mail-order service through which anyone could submit requests for copies of their catalog cards. Demand proved slow at first. They fielded only twenty-one such requests in 1896, but that number tripled the next year, and grew steadily for years to come; by 1912 the organization was fielding 1,500 requests per year.[24] The organization that the two men envisioned was to be more than just a library cataloging service, however. It would also incorporate bureaus responsible for coordinating publishing standards and other international initiatives in domains as diverse as statistics, teaching standards, patents, botany, and hygiene. The IIB's most important and ambitious initiative, however, would be the creation of the Universal Bibliography, which would be published under the Latinate name *Bibliographia Universalis*.

While the project continued to face resistance from foreign librarians, it met with a much warmer reception from the scientific community, many of whom were quick to recognize the advantages of universal indexing standards and were more receptive to the idea of international cooperation. Given the frequency with which scholarly books and papers were starting to circulate across national boundaries, many scientists immediately recognized the advantages of a system that removed the need to translate subject nomenclature across many different languages. Awash as they were in an ever-rising deluge of scholarly publications and accustomed to gathering in international congresses to establish consensual standards, several prominent scientists soon embraced the IIB's quest to produce a Universal Bibliography.

Otlet and La Fontaine found an enthusiastic early collaborator in Charles Richet, editor of the influential *Revue scientifique*—roughly the equivalent in its day to *Scientific American*. He began putting decimal classification numbers on every article that appeared in the journal. "Any analytical classification can be made only in an international language," he wrote, "and the only universal international language one could adopt is the language of numbers."[25] The project also attracted the support of Herbert Field, an American zoologist based in Switzerland, who managed to convince that nation's government to lend support to his *Concilium Bibliographicum*, an attempt to gather and publish bibliographic data about the most important literature published in zoology and related life sciences. Field hoped to publish these bibliographies in rapid succession, to help researchers stay abreast of the field. After corresponding with Otlet and La Fontaine, he embraced their approach to issuing bibliographies on index cards. Field agreed to publish his entries on the same card stock as his Belgian collaborators, and they soon began to collaborate on revisions to the classification.[26]

The IIB secured an important institutional beachhead when the French Association for the Advancement of Sciences agreed to join in a partnership, giving Otlet and La Fontaine hope that their universal classification scheme might find traction with the broader scientific establishment. The International Congress of Publishers soon followed suit, as did the International Union of Photography. Not all collaborations went smoothly. One French editor named Charles Limousin "became intoxicated with classification-building," as Rayward puts it, "and departed happily but wildly from the original scheme."[27] Otlet quickly put a stop to his efforts, refusing to allow a lone-wolf bibliographer to compromise the integrity of the system. From this point forward, he and La Fontaine insisted on managing the entire enterprise, down to an excruciatingly precise level of detail, reviewing and hand-correcting every submission from their network of scholarly partners. That dedication to the principles of consistency and standards is the hallmark of a good cataloger, and equally indicative of a controlling and inflexible personality.

At their next conference in 1897, Otlet described the project in characteristically sweeping terms as "an inventory of all that has been written at all times, in all languages, and on all subjects."[28] By this time, the IIB had started to produce subject-specific bibliographies on topics ranging from medicine and chemistry to Belgian women and Hungarian economics. Some of these works were original productions; others were adaptations of previously published works. The office also provided copies of any manuscript requested by letter, with a nominal fee for copying costs (this facility would be extended to telegraph requests in the years to come). By the end of that year, the collection consisted of 1.5 million cards.

Otlet hoped one day to distribute copies of the Universal Bibliography internationally, to make it a core reference source for every

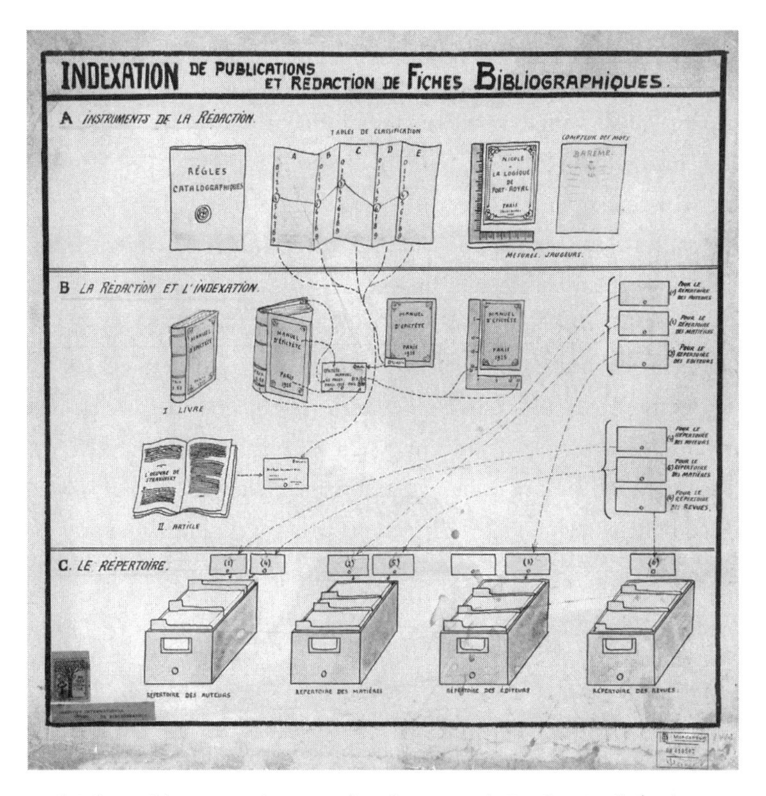

Paul Otlet, Publication Indexing and Redaction with Catalog Cards (*Indexation de Publications et Redaction de Fiches Bibliographiques*). Reproduced with permission from the Mundaneum.

major library in the world. He envisioned running it as a subscription service, with regular updates distributed by mail to its patron institutions. Eventually, he imagined a network of bibliographical agencies in every country of the world, each contributing to the collection and coordinating their efforts through a tightly managed process of standards and process controls. However, Otlet's aspirations far outstripped the technological reality of his time. There was as yet no reliable mechanism for copying such large numbers of index cards other than typing them out by hand (Otlet and La Fontaine offered a cash prize of 500 francs to anyone who could design a machine capable of reproducing index cards in large numbers). Eventually they invested the considerable sum of 22,000 francs in a costly printing scheme that never bore fruit. Instead, the IIB ended up distributing copies of the bibliography in book form, with a set of instructions directing subscribers to take out their scissors, cut up entries by hand, and paste them one-by-one onto index cards. Even four centuries later, Gessner's indexing technique was proving remarkably durable.

As the size of the collection increased, the conceptual problems of building a global classification scheme proved increasingly thorny. The scope of the undertaking—cataloging literally all the world's documents—demanded a similarly expansive system. Otlet and La

The Universal Bibliography, c. 1906. Reproduced with permission from the Mundaneum.

Fontaine became convinced that none of the currently available library systems—including Dewey's—would prove entirely up to the job; they would have to invent their own.

In 1892, Otlet published a far-sighted essay, "Something about Bibliography," in which he began to elaborate the thoughts that he had been refining since his adolescent days in the school library, when he first contemplated the power of the library catalog. These ideas finally seem to have crystallized in a few short pages, in which he laid out the basic framework of ideas that would guide much of his professional life to come. Setting his sights squarely on the problem of information overload, he lamented "the debasement of all kinds of publication," citing a proliferation of books, pamphlets, and periodicals that, he argued, should be "alarming to those who are concerned about quality rather than quantity." Echoing the complaints of generations of philosophers from Plato to Descartes, he fretted about the impossibility of keeping up with the spiraling output of published human knowledge. "If one spends a very little time reading these works, it seems that everything has been said, that everything is known and that further reading is pointless."[29]

Evoking the positivist spirit, he went on to hold out the possibility of a new philosophical unity rooted in a grand synthesis of human knowledge. Given the daunting volume of published information in the world, the situation seemed to call for a fundamental rethinking of the bibliographic arts. Otlet argued that most librarians had focused too much on the problem of guiding readers to a specific author or title on the shelf, with little regard to the actual content contained inside those books. He proposed a radically new approach: freeing information from the physical confines of the book. "What does the title of a book and the name of its author conceal?" he asked. In an age of burgeoning information, scholars would need methods of distilling information gleaned from across an ever-growing range of sources. To that end, he proposed a simple strategy, premised on the notion that all human knowledge could ultimately be broken down into four fundamental elements:

Facts
Interpretations of Facts
Statistics
Sources

Reduced down into these component parts, data could then be rearranged and interpolated into any number of new combinations, ultimately yielding what he called "a kind of artificial brain." The key to this approach would involve focusing the efforts of librarians and bibliographers on extracting the essence of what could be found between the covers. "The ideal," he wrote, "would be to strip each article or each chapter in a book of whatever is a matter of fine language or repetition or padding and to collect separately on cards whatever is new and adds to knowledge."[30] Otlet's willingness to dismiss the individual author's voice—mere "fine language"—anticipates the modern search engine, whose algorithmic purpose is to extract individual units of information from a vast collection of texts. Umberto Eco's distinction between "books to be read" and "books to be consulted"[31] seems relevant. The former category consists of novels, poems, essays, and other written works predicated on the bond of sustained attention that unites readers with writers, while the latter comprises the endless volumes of reference works, scientific documents, and other agglomerations of facts that seem well suited to the kinds of brute-force indexing that Otlet advocated.

Drawing on his Gessnerian experiments with filing systems, Otlet embraced the index card as the superior vehicle of information technology, arguing that such atomized information storage

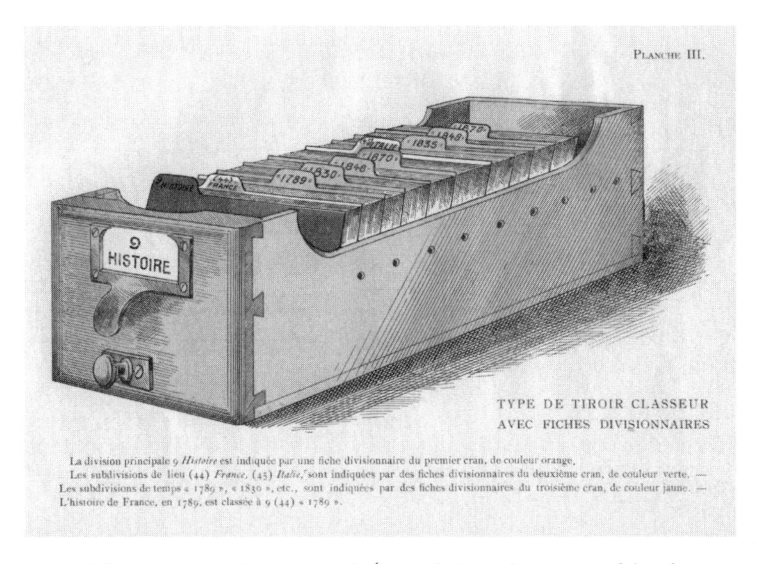

Type of filing drawer with index cards (*Type de tiroir classeur avec fiches divisionnaires*). Reproduced with permission from the Mundaneum.

would allow information to be broken into discrete units. "These cards, minutely subdivided, each one annotated as to the genus and species of the information they contain . . . could then be accurately placed in a general alphabetical catalogue updated each day."[32]

To accomplish that task, he proposed a "detailed synoptic outline of knowledge" that would disentangle facts, interpretations, statistics, and sources from the volumes that contained them, allowing each unit of information to be indexed, manipulated, and republished in new and evolving forms. "Such are a few of the ideas that come to mind when, after some investigation, we ponder what Bibliography and collective endeavor could achieve in advancing the social sciences."[33] Working from the germ of these insights, Otlet would spend the next five decades trying to realize that vision.

As work on the Brussels Classification progressed, Otlet and La Fontaine began to experiment with new methods for making it more powerful, so that it could move beyond the one-dimensional hierarchical classification of the increasingly popular Dewey Decimal System. While Otlet admired the clean logic of Dewey's classification and hoped to adapt it as the basis for their own, he aspired to create a far more complex, multidimensional system that would allow for deeply focused indexing and cross-referencing of related topics. Although the Dewey system had its flaws, it also featured several characteristics that Otlet found appealing. It offered a comprehensive array of subjects; used the universal language of numbers for identifying resources; and, perhaps most important, it was widely distributed and accepted as a de facto cataloging standard. If Europe and North America could converge on a common classification, Otlet hoped, the dream of a Comtean synthesis of all human knowledge seemed within reach.

In 1896, Otlet published a paper entitled "On the Structure of Classification Numbers," a dense but seminal piece of work in which he probed the limitations of the Dewey system and tentatively suggested an alternative. Whereas traditional library cataloging systems had focused largely on the problem of guiding a reader to a particular book, Otlet hoped to develop a far more inclusive, "universal" system: one capable of accommodating knowledge in multiple forms—text, photograph, chart, or recording—and at multiple levels of meaning, from book to chapter, to paragraph, to sentence, and on down to the level of irreducible "fact." To that end, he started to outline the contours of a classification system capable of capturing the full spectrum of human thought and expression, one that "should present a framework in which ideas are subordinated successively and in different ways," describing a future system that could one day become—shades of John Wilkins—"a veritable universal language."[34]

Otlet and La Fontaine wrote to Dewey to request permission to translate his classification into French and German. Dewey agreed. In the process, they also began to elaborate on the taxonomy to accommodate a widening sphere of subjects. The men eventually dubbed their cataloging system the Universal Decimal Classification (UDC). Following the Dewey system, the UDC divided all knowledge into ten categories, which in turn were divided further into a theoretically endless succession of subcategories. The UDC relied on a system of tables that outlined the conceptual structure of the classification. The main tables (or "schedules") contained top-level domains and their subordinate categories, namely:

0 Generalities
1 Philosophy; Psychology
2 Religion; Theology
3 Social Sciences
4 [Vacant]
5 Natural Sciences; Mathematics
6 Technology
7 The Arts
8 Language; Linguistics; Literature
9 Geography; Biography; History

The UDC's top-level subject classification scheme closely resembled the Dewey system, with a few notable exceptions: Language and Literature were merged into a single category, leaving one top-level category blank for future use. The UDC also dispensed with Dewey's three-digit minimu, thus 310 in the Dewey system would equate to 31 in the UDC. These differences notwithstanding, the UDC hewed closely to the Dewey system in terms of its top-level ontology. However, the UDC also included a second set of so-called common subdivisions that could be applied to all subjects; these included form, time, relation, proper name, place, and language. Another set of "special auxiliaries" included highly specific relationships suited

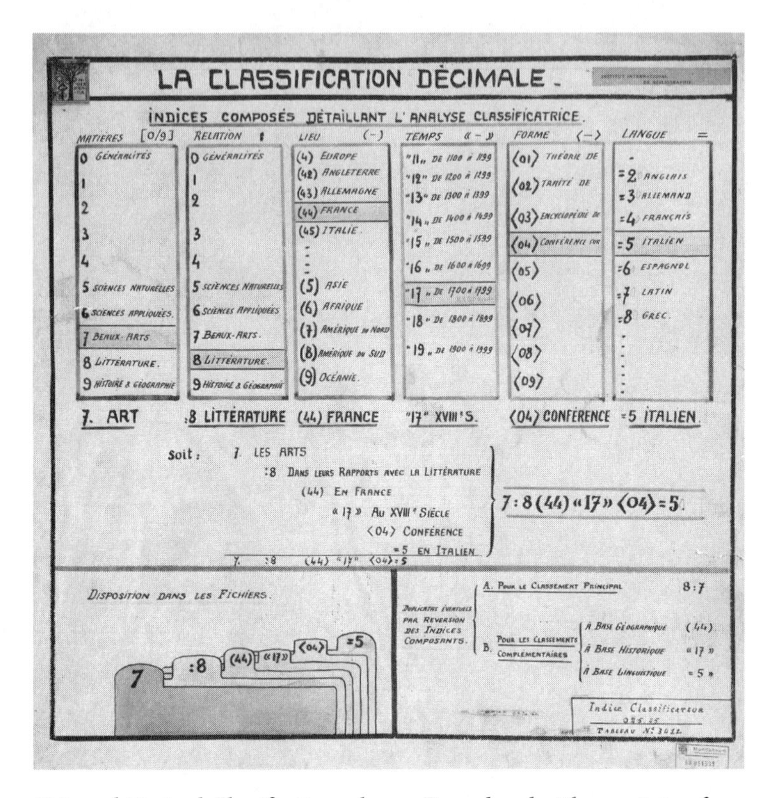

Universal Decimal Classification schema. Reproduced with permission from the Mundaneum.

to a particular domain. This innovation amounted to an incremental improvement over Dewey's system. The real breakthrough came when Otlet and La Fontaine decided to eliminate a cornerstone of the Dewey Decimal System: namely, the decimal itself.

Whereas the Dewey system relied on decimal points to separate numbers, the UDC featured a more flexible series of notations that would allow for the expression of more complex relationships between subjects. Beyond mere decimals, it allowed for a much broader range of semantic relationships than any previous library cataloging system. Using single characters, such as the equal sign, plus sign, colon, quotation marks, and parentheses, Otlet and La Fontaine created a symbolic language to encode relationships from one topic to another. For example, the plus sign would allow catalogers to assign multiple topics to a single source; colons allowed them to combine classification numbers; additional qualifiers allowed the use of parentheses to denote place and the equal sign to denote language; and quotation marks denoted time periods.

For example:
51 + 53 = Mathematics and chemistry
63:30 *or* **31:63** = Statistics relating to agriculture
339.5 (410/44) = Trade relations between the United Kingdom and France[35]
94(410)"19"(075) = History of the United Kingdom in the 20th Century Textbook[36]

This ability to create symbolic links between multiple topics marked a major conceptual breakthrough, allowing for a theoretically endless series of permutations that could express any number of semantic relationships. It also provided a powerful tool for indexing, permitting highly specific references drilling deep into the contents of books and articles. This ability to create symbolic linkages between entities and interpolate them with one another anticipates the logic of the modern relational database. At the time, however, it found expression in drawer upon drawer of index cards stored in the IIB's offices.

Previous library cataloging schemes, dating all the way back to the library at Alexandria, had relied on a linear list of single-subject headings. While a cataloger might choose to designate a book as covering several subjects, no one had ever proposed a mechanism for creating direct links between them. By the late nineteenth century, however, the rapid proliferation of published works had triggered a commensurate proliferation of subject headings, making linear cataloging systems increasingly unwieldy. Moreover, the rising tide of scientific and technical literature demanded more precise techniques for describing highly specific subject matter. Highly precise classification, coupled with the ability to "zoom out" on any subject by telescoping back through a series of numbers and expressions, allowed for a level of detail that would, in theory, allow the system to penetrate deep into the granular components of a book or article.

In practical terms, Otlet initially envisioned this as a physical volume that would consist of loose-leaf pages, constantly updated by expert authors collecting and extrapolating information from a wide range of primary sources. A federation of knowledge workers around the globe would develop expertise in particular domains and assume responsibility, acting as institutional gatekeepers to knowledge produced within each area. These experts would evaluate each newly published document, assess its content to determine whatever new facts might be contained inside, then transcribe that data onto an index card and assign a code drawn from the Universal Decimal Classification. Thanks to the linked structure of the UDC, each fact would thus take its place in relation to the universe of other known facts. Such a system, Otlet wrote, "allows us to find a place for each idea, for each thing and consequently, for each book, article, or document.... Thus it allows us to take our bearings in the midst of the sources of knowledge, just as the system of geographic coordinates allows us to find our bearings on land or sea."[37] Over time, he imagined, these Universal Books would eventually yield a World Encyclopedia, constantly updated with new information from the latest works as they were published, a kind of living master catalog that would encompass the totality of human knowledge.

By the time of the Paris Expo in 1900, the Universal Bibliography had grown dramatically, numbering more than 3 million entries stored on individual index cards. More than 300 members had joined the IIB, including scholars and organizations from England, France, Germany, the United States, and Bohemia, among dozens of other countries. The IIB still encountered resistance from several quarters of the bibliographical establishment, however, most notably the Royal Society in London, which had started to promote its own original classification for the sciences. But by this time Otlet and La Fontaine had built up enough international credibility that they managed to secure a spot in the Expo's Grand Pavilion.

In addition to the permanent exhibit of the Universal Bibliography—which they themselves staffed, answering questions and promoting their vision to scholars and anyone else who would listen—they also convened a meeting of librarians and bibliographers, adding their voices to the Expo's ongoing caterwaul of international congresses. The conference set out to address the problems of creating universal "bibliographic repertories" that would satisfy scholars of all stripes. Ninety-one participants attended, including prominent librarians and bibliographers from all over Europe and North America (notably absent were representatives from England and Germany), including no less an eminence than Melvil Dewey. The delegates shared examples of their work and debated the finer points of bibliography, copyright law, and the need for better statistics about the volume of printed material published each year.[38]

Discussion of such a high-minded scholarly endeavor was not at all out of place at the Expo. The rise of international scholarly associations had brought the problem of bibliographical standards into sharp focus, and the topic would doubtless have held interest for a great many attendees. Whereas once individual libraries had contented themselves with maintaining their own ad hoc classification schemes and indexing systems, the growing international collaboration among scholars—especially in the sciences—created a need for more consistent methods of describing the subject of any given book or article and framing that subject in a larger scholarly context that mapped the conceptual relationships between increasingly specialized disciplines and fields of study. Otlet felt strongly that for human knowledge to progress in this increasingly networked environment of international associations, scholars would have to agree on a consensus understanding of how their fields related to each other. "All sciences are auxiliary to each other," he wrote, "and owe the best of their progress to their connections."[39]

Nonetheless, as with many academic conferences, a schism emerged early on, pitting advocates of traditional bibliography—those who advocated highly selective, humanistic approaches to curating human knowledge—against the so-called scientific bibliographers, who believed in the possibility of more inclusive, systematic, even universal schemes. Otlet fell squarely into the latter camp. He used the conference as a platform to advocate his belief in international standards, arguing that librarians should embrace common protocols, including a universal set of subject headings that could enable them to share their work more effectively across institutional and national boundaries. Although the conference ultimately resulted in a series of platitudinous proclamations (as conferences tend to do) recognizing the value of different approaches, Otlet had nonetheless seized the opportunity to establish himself as an influential voice in a growing international conversation.

When the Expo closed on November 12, 1900, Otlet and La Fontaine walked away with a Grand Prize.

4

The Microphotic Book

The telegraph was only four years old in 1848, when the Associated Press first went into operation, bringing with it a cavalcade of "news from nowhere, addressed to no one in particular,"[1] as Neil Postman once put it. While the first recognizable newspapers had started to pop up in Europe beginning in the seventeenth century, not until the nineteenth century did they start to circulate widely throughout Europe and North America. The popular appetite for newspapers grew in tandem with a rapid decline in production costs, thanks to innovations such as steam-powered printing presses and cheap rag paper. Otto Mergenthaler's invention of the Linotype in 1884 drove down publishing costs even further, enabling the papers to expand their contents dramatically. Until then, most newspapers published at most eight pages in weekly editions, painstakingly set by hand compositors plucking letters, ingot by ingot, from a typecase divided into upper and lower cases. By the turn of the twentieth century, many of those old "stick men" (and more than a few women) had lost their jobs to a new generation of Linotype-equipped press rooms with lower labor costs and higher productivity, regularly turning out daily editions of sixteen or thirty-two pages or more.

These industrial printing technologies allowed information to flow more easily from editor to reader; they also allowed it to flow from paper to paper. Thanks to the telegraph and an increasingly efficient postal system, late nineteenth-century newspapers borrowed freely from each other (with and without attribution) and shared prewritten material via news syndicates like the new Associated Press wire service.

Journalist Tom Standage argues that the spread of the nineteenth-century telegraph network bears more than a passing resemblance to the present-day Internet, both in its distributed technical architecture and in its social and cultural effects.[2] The early telegraph pioneers saw the new invention as far more than just a mechanism for speeding up the transmission of text. "If the presence of electricity can be made visible," wrote the telegraph's inventor, Samuel Morse, in 1837, "I see no reason why intelligence might not be instantaneously transmitted by electricity to any distance."[3] Both Alexander Graham Bell and Thomas Edison worked on telegraph-based devices that would enable photographs to be transmitted in electric form, by translating an image into a collection of binary dots that could be encoded on one machine and reassembled on the other—something like a fax machine.

One American newspaperman, Walt Whitman, recognized the world-changing potential of the new technology. He first encountered a telegraph machine in 1848, while working as an editor at the *Brooklyn Daily Eagle*, where he noted that the newspaper had successfully transmitted and received the governor's 5,000-word message, received at noon and "in type, printed and for sale in Brooklyn and New York by 4 o'clock!"[4]

Years later, the telegraph would serve as an essential reference point for Whitman's masterwork, *Leaves of Grass*, in which he celebrates the "seas inlaid with eloquent gentle wires" that gave passage to "the word En-Masse." He now saw the device as an apt metaphor for his own poetic vision of a more unified world, invoking the technological wonders of the age as both metaphor and manifestation of his hope for a more perfect Western Union. "See, the electric telegraph stretching across the continent," he writes. "See, through Atlantica's depths pulses American Europe reaching, pulses of Europe duly returned." Looking ahead hopefully to a world where old enmities are finally healed, thanks in part to the advent of new communications technologies that might reduce the physical and emotional distance between people, he imagines a future in which the liberating currents of democracy ripple around the globe. "I will make divine magnetic lands."[5]

In 1898, another former newspaperman, Mark Twain, published a short piece of speculative fiction called "From 'The London Times' of 1904." Here he introduces a hypothetical device called the Telectroscope, capable of transmitting sound and images via cable all over the world. "The improved 'limitless-distance' telephone was presently introduced," writes the narrator, "and the daily doings of the globe made visible to everybody, and audibly discussable too, by witnesses separated by any number of leagues."[6]

Twain himself was no stranger to technology, having bankrupted himself on a failed experiment in automated typesetting called the Paige Compositor, a device with 18,000 moving parts that purported to set type 60 percent faster than the Linotype. Twain remained fascinated by the possibilities of the new technologies, however, maintaining a close friendship with electricity pioneer Nikola Tesla and even patenting a few devices himself, including a replacement for suspenders and a self-pasting scrapbook. In his writing, he had explored new technologies like fingerprinting (in *Pudd'nhead Wilson*), and even probed the possibilities of time travel (in *A Connecticut Yankee in King Arthur's Court*).

Twain's business acumen may not have matched his literary abilities, but that he could envision something like an electric telescope at the beginning of the twentieth century suggests that the basic contours of the modern electronic information ecosystem were already coming into view. The rise of the telegraph, Linotype machines, and photography all seemed to point toward a new kind of media landscape: fluid, participatory, and potentially overwhelming. The emerging mass media had already started to shift the economics of knowledge production, creating a new class of professional writers and editors trying to cater to the interests of a large and increasingly literate general public.

In Belgium, newspapers had started to proliferate as well, with daily papers in every major city. By the early 1900s, the kingdom's news publishers had formed the Belgian Union of the Periodical Press. In 1906, its members elected a new vice president: Paul Otlet.

By this time, Otlet had emerged fully into the public eye. The withdrawn youth and failed lawyer had finally found his footing in the world and assumed a quasi-statesmanlike role in his capacity as head of the IIB. Otlet had started to take on a more public persona, even representing the Belgian government at major international gatherings. Still only twenty-nine, he seemed to enjoy playing the role of young eminence, using his position to ingratiate himself with a number of other organizations and institutions. Otlet also leaned heavily on his partner Senator La Fontaine's political connections.

Otlet's growing involvement with governments and international associations also brought him into closer contact with the publishing industry: printers, photographers, lithographers, bookbinders, and other artisans of the local book trade. These interactions were not always positive—the IIB perpetually had trouble paying its printer, dependent as it was on the timing of its government subsidy. Two years earlier, Otlet had founded a Museum for the Book, graciously agreeing to serve as its first president. Located near the Royal Library and Museum on the Rue de la Villa Hermosa in Brussels, the museum served as a professional gathering place as well as a public exhibition space with classrooms and a lecture hall outfitted with a lantern projector and screen.[7] Building on his success with the Museum of the Book and his new role as president of the newspaper association, Otlet saw an opportunity to apply his ideas about cataloging and classification to the world of periodicals, by creating a similar hub for the international newspaper industry. He hoped that such an institution could serve as not just an exhibition hall, but also as a central clearinghouse for all the world's newspapers. He began directing his colleagues at the IIB to put extra effort into expanding its collections of newspapers and journals and initiated plans for an international newspaper museum whose charter would be to collect and curate periodicals from all over the world, to create "a collection as complete as possible of works on the Press as well as forming a complete collection of older periodicals and newspapers."[8] It would also collect works on newspaper history and key works on copyright law and freedom of expression. Soon enough, however, Otlet realized the enormity of that task. Early on he complained to the press association that "our personnel cannot devote all the time desirable to integrate the specimens into the collections which grow more and more."[9] He pleaded with them to give him more resources.

In the meantime, Otlet and La Fontaine continued to press forward with their schemes to unify the world's documentary practices. Drawing on their government connections, in 1905 the two men wrote a proposal for a Documentary Union of Governments, outlining how the bibliographic methods they had perfected with the Universal Bibliography could yield improvements in the way government organizations communicated, both internally and with one another. The growing flood of memos, reports, dossiers, and so forth seemed to call out for standardization that would make it easier to collect and collate their intellectual output by following the Universal Decimal Classification. Otlet began to envision a new Universal Repertory of Documentation, modeled after the Universal Bibliography, providing information about every document's author, country, subject, and date. Eventually they imagined this effort converging with similar efforts by the world's universities, scholarly associations, and other institutional entities, all ultimately taking shape as a vast network (*réseau*) of connected entities, with a new international administrative body at its core.[10]

With the support of the Belgian government, Otlet and La Fontaine sent invitations to thirty-five countries to join the proposed documentary union. Much to their dismay, however, all of

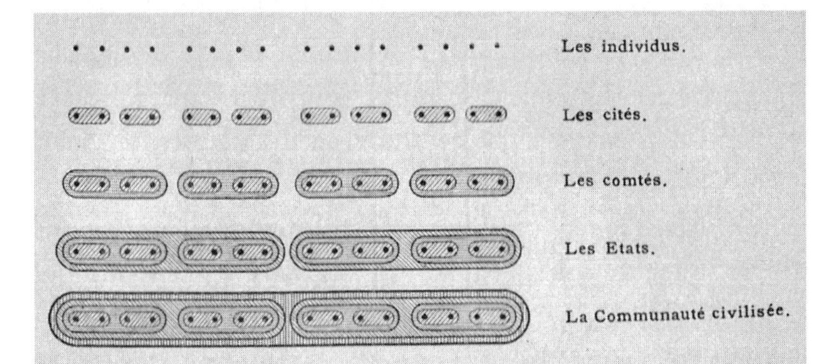

Progressive Expansion of Social Structures (*Extension progressive des structures sociales*). From Publication no. 25, Office Central des Associations Internationales (1912). Reproduced with permission from the Mundaneum.

the major world powers declined. England and France refused outright, as did Switzerland and Cuba. The American Library Association offered some encouragement, however, offering to lend its support to foster further discussion with the U.S. government. Only Persia and Denmark offered their unconditional support. Undeterred, Otlet and La Fontaine continued to press for their vision of making Brussels an international communications hub. While they recognized the difficulty of fostering agreement among the world's governments, they believed they might have better luck selling their idea to members of the growing number of international associations located in their own backyard.

Since the mid-nineteenth century, Belgium's history of neutrality—dating back to the 1839 Treaty of London—had made it an attractive gathering place for international associations. Brussels had long played host to international congresses, usually centered around a particular topic like welfare, statistics, or continental peace. These gatherings would often take place simultaneously, often in connection with large public exhibitions, festivals, or other events that would give the organizers the chance to entertain the attendees. They would throw banquets and soirées, exchange speeches, and confer medals, ribbons, and honorary titles upon one another. With Leopold's support, the city began to play host to a seemingly endless parade of international associations, like the International Institute of Photography, the International Congress for the Study of Polar Regions, the Congress of the Federation of Regional Hunting Societies, and the International Documentary Office of Fisheries. By the end of the decade Brussels would be hosting more international events than any other city: more than Paris, twice as many as London, and ten times as many as Berlin.[11]

When one of these groups convened a meeting, the working sessions typically featured a procession of speakers making highly

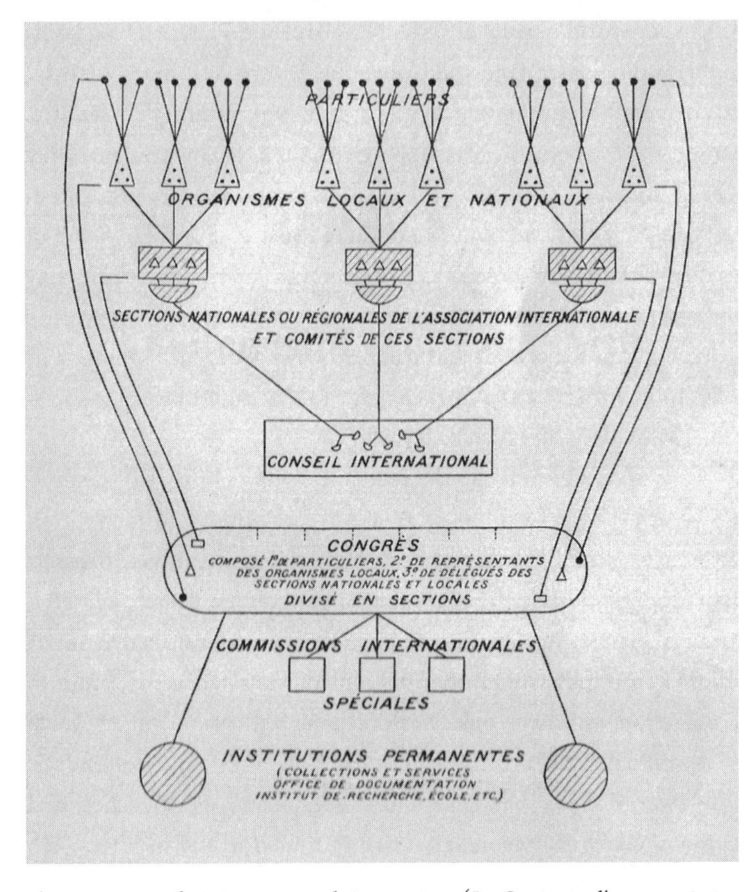

The Structure of an International Association (*La Structure d'une association internationale*). From Publication no. 25, Office Central des Associations Internationales, 1912. Reproduced with permission from the Mundaneum.

scripted remarks, all scrupulously documented and disseminated to the attendees. As a result, these once-informal societies began to take on a more active role as producers of written knowledge and institutional arbiters of fact.[12] Along the way, they left behind a growing paper trail of reports, documents, newspaper clippings, scholarly articles, and books—many of which found their way into

Otlet and La Fontaine's archive, where Otlet began to recognize the "living elements of a general organization." He envisioned a central role for the IIB in directing the flow of this information into a more cohesive whole, by giving it "encouragement and a system in order to emerge."[13]

Such an undertaking would require more than librarians. It would call for a new breed of professional, what Otlet called a "documentalist." Unlike the traditional librarian—whose tasks consisted primarily of collecting, archiving, and curating books—the documentalist would play a far more active role in analyzing and distributing recorded knowledge. Applying Otlet's monographic principle and following the cataloging rules of the UDC, the documentalist would collect, analyze, and summarize documents from multiple sources, then disseminate them into a larger apparatus of recorded knowledge. International associations would also be expected to hire their own documentalists to maintain their bibliographies, collecting information about their own published documents as well as other relevant materials in their fields, such as journal articles and scholarly monographs. A few years earlier, La Fontaine had published a source book of documents on international law, as well as a comprehensive bibliography on peace and international arbitration. That work would serve as a template for the kind of international bibliography that the men hoped to create on a broader scale.

Otlet envisioned a centralized office that would develop policies and procedures for international associations to follow in creating their documentation, with everyone involved "adhering to a common plan."[14] Governments would play a role as well, given their role in creating standards for exchanging information across national lines—responsibility analogous to the role they were beginning to play in upholding international copyright law. Each of these organizations would serve as a kind of filter, gathering data in the form of reports, books, photographs, and audio and visual recordings, with a team of documentalists on hand to coax that data into reusable form, ready to be absorbed into the universal catalog.

As the international associations continued to expand their presence in Brussels, Otlet and La Fontaine applied their unique combination of bibliographical skills and bureaucratic savvy, cultivating personal relationships with the leaders of many international associations. In 1906, they convened a meeting of representatives from several Brussels-based associations to discuss how they might work together toward "unification and progressive organization." That discussion quickly moved beyond a discussion of bibliographical standards, as the attendees began to explore the deeper organizational challenges inherent in trying to foster cooperation among the far-flung members of their associations. The meeting attendees began to ponder the possibility of a new international body, charged with promoting the shared interests of international associations by taking on a quasi-governmental role, acting in "the interests of the whole world—as though it was comprised of a single nation above individual nations."[15] That lofty sentiment went far beyond the question of organizing documents, pointing toward a more expansive vision of what such an organization might do. It also echoed the spirit of internationalism that was starting to percolate throughout Europe and North America, the same sentiment that would eventually give rise to the League of Nations (a term coined by the Frenchman Léon Bourgeouis in his 1910 book *Pour la société des nations*[16]). Indeed, Otlet and La Fontaine's work in organizing international associations would pave the way for them to play instrumental roles in the formation of the League a decade later.

In 1906, the notion of a functioning global federation still seemed a long way away, when in the aftermath of that meeting they created a new Central Office of International Institutions. The organization's first task was simply to compile a directory of all existing international institutions: surveying their history, structure, and the methods they used to create documents. By 1908 the Central Office had won the support of the Belgian government, which agreed to support their efforts by providing office space in the Mont des Arts and drafting a law to safeguard the status of international associations in Belgium. Otlet and La Fontaine gradually widened the scope of

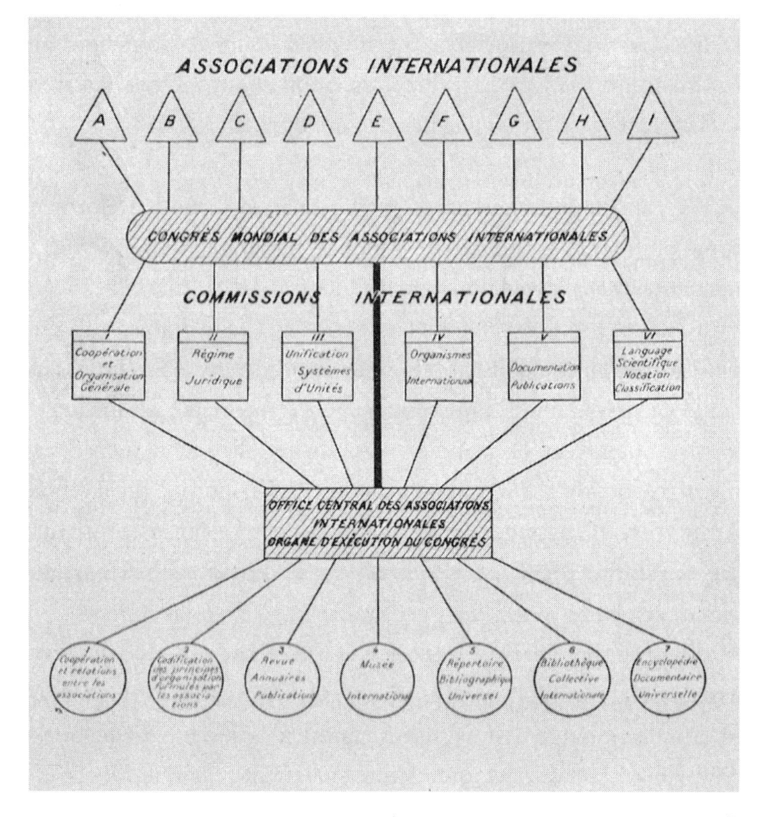

The Union of International Associations (*L'Union des associations internationales*). From Publication no. 25, Office Central des Associations Internationales (1912). Reproduced with permission from the Mundaneum.

their efforts, beginning to focus on legal issues and scientific and technical standards and creating standards to govern the exchange of documents between organizations.[17]

The challenge of archiving all these documents would require more than bureaucratic standards and practices. There was also the basic physical problem of finding enough space to house it all. The sheer heft of the world's written output—newspapers, documents, memos, research reports, and the rest—made it all but impossible to conceive of archiving everything in one place. Such an undertaking could never possibly scale, unless it could all, somehow, be shrunk into a more manageable form. Soon enough, Otlet began to set his sights on a promising new technology: microfilm.

In 1839, an Englishman named John Benjamin Dancer performed some of the first experiments with microphotography, perfecting a technique that allowed him to reproduce images of objects, reduced by a ratio of 160:1. By 1853, he had perfected the technology to the point where he could distribute copies of printed texts on slides that could be viewed under a microscope. A few years later, a French optician named René Dagron further perfected the technique, securing a patent for a microfilming process that allowed the French Army to send messages across German lines by means of carrier pigeons during the Franco-Prussian War of the 1870s; the messages could then be retrieved, enlarged, and printed by the recipients.

Despite these early advances, microfilm had failed to achieve commercial viability in the ensuing years. Most serious photographers continued to dismiss it as a gimmick—akin to writing one's name on a grain of rice: interesting, entertaining, but hardly practical. Otlet, however, recognized the enormous potential that microfilm held for storing the world's growing corpus of documents, and he began to explore its possibilities in the early 1900s. Working with an inventor named Robert Goldschmidt, he began to describe what he called the "microphotic book" that relied on microphotography to reproduce the images of individual pages that were then stored in a standard format identical to the size of an index card in the catalog, with each page of text occupying one square centimeter on the microfiche. He envisioned a specialized reading machine adapted from microscope and typewriter parts to move a magnifying lens over the film and project it onto a blank surface. He imagined that this technology would prove particularly useful for archiving out-of-print books, patent filings—and newspapers.[18]

Using the standard-sized microfiches, Otlet hoped to create an expanded version of his massive catalog, one in which the microfiche documents would sit side by side with their attendant card catalog entries. Thus, a researcher could peruse the catalog, consult an index card, then pull up the desired documents on the spot via microfiche—without having to retrieve a physical paper volume from a library shelf. The catalog itself would serve simultaneously as reference and source material, a kind of integrated information storage and retrieval system. "If one had the necessary resources at one's disposal," he wrote, "all of Human Thought could be held in a few hundred catalogue drawers, ready for diffusion and to respond to any request."[19]

Instead of printing books and journals on paper, Otlet envisioned that one day publishers would publish their contents directly onto index cards as "autonomous elements" that could slot neatly into the catalog, ready for future scholars to retrieve and reuse in new forms. In 1907, he persuaded the Belgian Sociological Society to produce one of its publications directly onto index cards. The catalog would become more than a repository; it would become an active tool for the production of knowledge.

Otlet had long since imagined that his universal knowledge repository could extend well beyond being a bibliographical catalog. As he continued to work on the Universal Decimal Classification, he realized that the system could encompass not only books but also periodicals, photographs, moving pictures, audio recordings, maps, or even museum collections—anything that could be described in terms of a set of subjects. He also began to expand the definition of what might constitute a "document." Today, most of us would likely associate the term with one or more pieces of paper. By the nineteenth century, however, that term had become increasingly anachronistic. Otlet proposed a far broader definition, suggesting that "document" could refer to any object manifesting any kind of graphic symbols—letters, numbers, images—captured in any form of media in order to express any form of human thought. Otlet eventually came to use the shorthand "book," anticipating an ever-expanding landscape of information in which almost any object can serve as a document.

By including a full range of possible media types, the catalog could far exceed its original role as a bibliography. In addition to newspapers, Otlet now began supplementing the collection with photographs, posters, postcards, and other forms of visual material. He hoped one day to transform them all into a universal encyclopedia of images, or a "museum of museums."[20] These objects could also take their place in microphotic form inside the catalog.

Ultimately, Otlet imagined all of these sources converging into what he called a Universal Book—an entity that would transcend the limitations of traditional physical formats to provide an endlessly extensible record of human knowledge in all of its forms. These elements together could one day coalesce as "paragraphs in one great book, the Universal Book, a colossal encyclopedia framed from all that has been published."[21] Years later, Otlet would build on his vision of microfilm to describe a "Document Super-Center," which would act as a central distribution point for microfilmed material that could be distributed via a universal network. "Gradually a classified *Microphotic Encyclopedia* would be formed from them," he wrote, "the first step towards new *microphotolibraries*."[22]

Like his Renaissance forebears, Otlet saw the ultimate synthesis of human knowledge taking shape in a single grand encyclopedic project. The Universal Bibliography was but the beginning. One day it would grow into an even more comprehensive resource that transcended the limitations of the written word and approached "the domain hitherto reserved for the universities of the spoken word." Someday, he imagined, there might be a grand "*Archivium* of humanity," and the Universal Bibliography would be its "true catalog."[23]

As Otlet's professional life flourished in the first years of the twentieth century, his family life fell apart. Burdened by his father's debts, Otlet's extended family still suffered from serious money problems; and Paul's work commitments limited his ability to intervene in their financial squabbles. When Édouard Otlet died from a heart attack in 1907, his oldest son went through a prolonged period of emotional turbulence, further exacerbated by a series of financial calamities. The elder Otlet's financial affairs had been in disarray for years, following an ill-fated investment in Spanish railway and mining companies. Financial problems would bedevil the family for years to come, forcing them gradually to sell off their other holdings to settle their debt. Though he came of age as an entrepreneur during the heyday of King Leopold when great fortunes were made, Édouard Otlet had never exhibited much aptitude for managing his money. He had, as Françoise Levie suggests, the soul of a poet. Or, as a prominent Belgian nobleman replied when his name came up in conversation, "Do not speak to me of Édouard Otlet. He is jinxed."[24]

Paul Otlet's single-minded dedication to his work was matched only by his wife Fernande's near-total lack of interest in it.[25] The two had long since drifted into separate orbits, spending less and less time together. In the wake of his father's death, with his family life in disarray, Paul started spending more time at his partner Henri La Fontaine's house. There he enjoyed the company of his friend and partner's sister, Léonie. Every Tuesday they would share dinner—a tradition that would last for much of the rest of his life. This friendship would never blossom into romance, however. Instead, he became infatuated with a friend of Léonie's, Cato Van Nederhesselt, a Dutch woman who exhibited far more interest in Otlet's work than Fernande. The two enjoyed a discreet dalliance, meeting in secret on a few occasions. Eventually, Otlet initiated divorce proceedings against Fernande. By 1909, the divorce was complete. Otlet wrote to Cato: "Now it is finished!... You have made me know the greatness of a loving heart."[26] Theirs was to be, by all accounts, a happy marriage.

Throughout this period, Otlet continued his close working relationship with La Fontaine. Between 1900 and 1910, the two worked side by side so often that it is sometimes difficult to determine where one man's work ends and the other begins. Their names appear together on numerous reports, letters, and official decrees. Rayward, for one, speculates that Otlet may well have been the primary mover during this period, drawing on La Fontaine's prestigious name while his partner occupied himself with his job in the Senate. Whatever the case, the men had started to move in the highest circles of Belgian political and intellectual life.

When King Leopold II died in 1909, he left a large personal fortune gleaned largely from his colonial adventures in the Congo. The disposition of that fortune remained uncertain; Otlet and La Fontaine hoped to persuade his estate to devote at least some of it to their grand plans for transforming Brussels into a world capital of intellectual endeavor.[27] Leopold's spirit of international expansion also lived on in the 1910 Universal Exposition of Brussels, where Otlet and La Fontaine convened the World Congress of International Associations (as it was now called), attracting delegates from 137 countries, as well as 13 governments and 4,000 individuals. Nobel laureate Wilhelm Ostwald took part, as did the great Belgian chemist Ernest Solvay, the French prince Roland Bonaparte, and Brussels mayor Adolphe Max.

The World Congress set about creating a bureaucratic infrastructure to accomplish its mission: establishing rules and procedures and standards and protocols, and passing a series of resolutions intended to govern their cooperation going forward. The Congress endorsed the notion of an international bibliographical standard, with the attendees giving their assent to the notion that "all information about bibliography and documentation should be coordinated, and a distinct brand of study created."[28] Further, the conference endorsed standardization of the 75 x 125 mm catalog card and the unification of the Dewey Decimal System with the Universal Decimal Classification. Most important, the Congress called for a permanent headquarters to be established in Brussels, along with "a super-national statute for non-profit organizations" and a central office for legal and contractual work: the Union of International Associations. Finally, Otlet and La Fontaine had fulfilled their vision of a central administrative body to coordinate the work of international associations.

Throughout the period from 1900 to 1910, Otlet and La Fontaine continued to strengthen their efforts to build a universal classification scheme capable of handling all the world's published information. To that end, they continued to maintain a regular correspondence with Melvil Dewey about the development of the Universal Decimal Classification, which they hoped would serve as a conceptual underpinning for all of their projects. The utopian vision they were starting to entertain would hinge on the availability of a common, universally recognized classification system. It was their fondest hope that the Dewey and UDC systems would continue to remain in lockstep with each other. For the first few years the Belgians and Americans collaborated regularly, exchanging and reviewing proposed updates to the classification tables. Dewey initially responded supportively, vowing to do his best to keep the subject classifications in sync between the two systems. "We feel strongly the need of having the new edition in harmony with your work," he wrote to Otlet, going on to praise his "rare sympathy, skill and efficiency."[29]

As the years progressed, however, Otlet and La Fontaine started to propose their own revisions and expansions to the Dewey system, often in concert with some of the groups involved in the Union of International Associations. Eventually, the two systems began to drift apart. For example, Otlet found that the Dewey scheme for mathematics failed to accommodate a set of new headings recently developed at the International Congress of Mathematicians; another of Otlet's collaborators felt that the American chemistry tables seemed to reflect an overreliance on one particular text. But discrepancies between the two systems continued to creep in nonetheless. La Fontaine, reviewing a set of proposed changes to the Military Science tables, asked whether the Americans would consider adopting UDC's punctuation methods (for example, the Dewey notation 355.342.1 would translate to 35.534.21 in the UDC). This method, he argued, would allow for more fine-tuned categorizations, in part by allowing for new uses of zeros and double zeros to create new categories. Otlet would later ask the Americans to consider incorporating the UDC's revised numbering standards and punctuation marks. Their request at first met with no response. Otlet persisted, and eventually Dewey delegated the correspondence to his assistant, May Seymour, who wrote to inform Otlet that they would not include them in the next edition. "We admire greatly the ingenuity of the IIB combining symbols and appreciate their convenience," she wrote, before going on to explain that they would not be using them, citing a concern that the new system "makes the numbers look so perplexingly complicated as to prejudice many persons at first glance beyond the power of argument." She also worried over the costs that would be associated with forcing American libraries to adopt the UDC methods. Moreover, she was apprehensive about the ability of library staffers to work with the numbers because "libraries have to use such cheap help to get books from the shelves and replace them that complicated numbers cause many mistakes."[30] Predicated as it was on an industrial model of standardization and economies of scale, the Dewey system was meant to be cheap, simple, and easy to implement. There would be no room for the nuance and sophistication of the Belgians' enhanced system.[31]

On receiving the message from Dewey's office, Otlet felt keen disappointment. He was concerned that a break with Dewey and the Americans would doom the UDC to the status of historical also-ran. "I observe with great regret," he wrote, "numerous divergencies in the classification."[32]

5

The Index Museum

In 1892, a Scotsman named Patrick Geddes bought the Short's Observatory, a popular attraction in downtown Edinburgh that consisted of a two-story observation tower set atop a seventeenth-century townhouse, featuring commanding views of the city and the nearby Forth Valley. The previous owner, Maria Short, had inherited a large collection of telescopes from her father, an astronomer and optician. After acquiring the building from its previous owners, she outfitted the rooftop with a large Camera Obscura, an optical remote-viewing device whose origins date back to ancient Greece. The camera consisted of a periscope and a series of mirrors that reflected living images of the outside world onto a circular table in the center of a darkened room. Using the camera, observers could enjoy close-up views of the city and surrounding areas from high above—an experience not far removed from watching a live television feed (or perhaps a more timely analogy might be a webcam). Such real-time voyeurism is unremarkable to us today, but in 1892 the prospect of streaming live images onto a remote projection screen probably seemed like a kind of conjuring act. For years, the camera had proved a popular attraction for Edinburgh residents and visitors alike.[1]

Geddes purchased the tower in hopes of using the camera as a popular draw for a more ambitious experiment: a first of its kind "sociological laboratory." Though originally trained as a biologist, he had taken a keen interest in the emerging field of sociology, where he applied the principles of the scientific method to the study of human societies, promoting the importance of careful observation in exploring the interplay between people, their work, and the places in which they lived. Eventually he would make important contributions to the field of urban planning, advocating a human-centered approach to building that stood in stark contrast to the centralized urban grid planning that had gained so much traction in cities like New York. In 1892, he was still developing his ideas, while conducting minutely detailed surveys of his native Edinburgh.

Geddes had developed a keen interest in the problem of Edinburgh's slums, which were rapidly expanding as the city population grew. From high atop his new building, which he renamed the Outlook Tower, he saw the city he loved becoming engulfed in cheap housing, caked in dirt, choked by smoke, and stewing in raw sewage. But most Edinburgh residents steered clear of neighborhoods like the infamous Leith slums, a district of "squalid lanes and closes" that few people of means had any reason to visit. The city's ruling classes tended to dismiss the newly urbanized residents as lazy and unkempt people who were quickly ruining the quality of city life.[2] Geddes, like other civic-minded reformers, took a broader view of the problem, arguing that the lives of all Edinburgh residents were inexorably intertwined, whether they liked it or not, and that the solution to the slum problem involved taking a more integrative and constructive view of city life. Geddes hoped to educate the public about a reality that many of them seemed disposed to ignore and perhaps incite them to action. By using the Camera Obscura, along with a series of exhibits about the city and its environs, he intended to cultivate a broader understanding of the problems facing Edinburgh and the world at large, by giving the public—literally—a bird's-eye view.

Visitors to the Outlook Tower would begin in the Camera Obscura, where they would encounter live projected images of the city streets and nearby farms that encouraged them to contemplate the interrelationship between the life of the city and the surrounding countryside. They would then wind their way down through successive floors of the building, each featuring a display intended not only to inform them about the city itself, but also to broaden their perspective on the relationship between the city and a wider, increasingly interconnected world. "The general principle is the synoptic one, of seeking as far as may be to recognise and utilise all points of view—and so to be preparing for the Encyclopaedia Civica of the future," Geddes wrote when describing the Outlook Tower in his 1924 book *Cities in Evolution*.[3] In the foyer outside the camera room, Geddes installed stained-glass windows devoted to a wide range of topics, such as botany and zoology—glowing triptychs laden with visual data. On the floor below was an exhibit about the history of Edinburgh; on the floor below that, the subject was Scotland, followed by Europe, and then, finally, the world. On the ground floor was a darkened chamber outfitted with a single chair: the Inlook Room, in which each visitor could reflect on what he or she had just learned.

Geddes's assistant and eventual biographer, Amelia Defries, described her first experience of touring the museum in his company: "As I stood, crammed with new knowledge, and upset by such sudden change in my ordinary ways of looking at things, Geddes pulled aside a curtain. 'In here you may rest awhile.'" She proceeded to sit alone in the rough-walled room, where she recalled "turning over in memory the outlook and its mirrored reflection, and the particular detailed studies of these. But I came to feel its emotional reaction. Here is the place where one's picture is conceived—not copied—from Nature. It is the room of the Weaving of Dreams."[4]

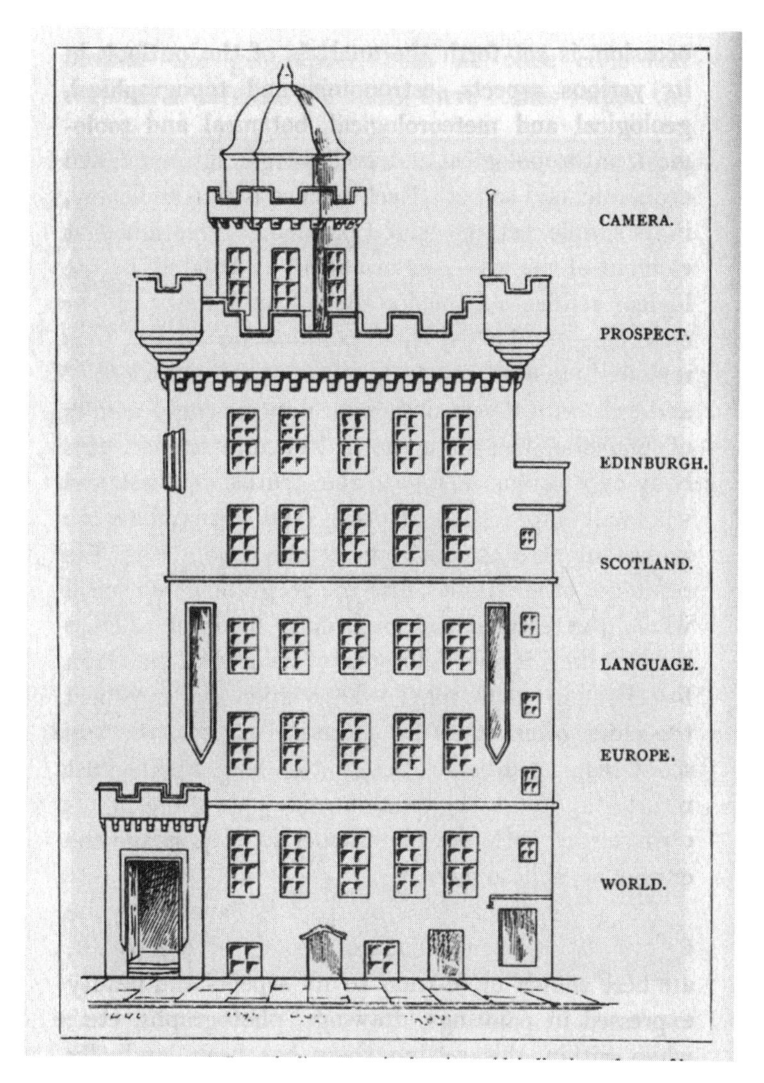

"Outlook Tower in diagrammatic elevation, with indications of uses of its storeys." From Patrick Geddes, *Cities in Evolution*, 324. Image courtesy of Chris Renwick.

Geddes's dreams of a transformed city stood in stark contrast to the gritty reality of Edinburgh, where a massive influx of new arrivals from the country had left the city contending with desperate poverty, high infant mortality rates, and widespread illiteracy. These social ills reflected not just the socioeconomic travails of that particular region but a much larger problem, involving the fragmentation of human perspectives throughout the industrialized West. Geddes did not see industrialism as inherently poisonous, however, nor did he advocate a Luddite return to a rural lifestyle. Rather, he envisioned a postindustrial city, imagining that one day even drab Edinburgh might transform itself into a "city of happy and healthy artists," creating wealth and prosperity for all, and living in concert with the surrounding countryside.[5] The display in the tower would play a critical role in this transformation, functioning as an active force for social change rather than a mere popular attraction.

Geddes's background in the life sciences equipped him with a strong interest in the problem of taxonomy, the classification of life forms into categories. As he widened his circle of interest into other realms, including sociology and urban planning, Geddes brought that same disposition to bear, trying to establish a "a colossal balance-sheet," as he put it, that would "harmonise the labours of all the schools."[6] Geddes deplored the tendency of academic disciplines to splinter—what Max Weber famously called the "iron cage" of specialization. The antidote, Geddes believed, was to build bridges through synthesis, creative classification, and educational outreach. "The teacher's outlook should include all viewpoints," he said in his farewell address to his students at the University of Dundee. "Hence we must cease to think merely in terms of separated departments and faculties and must relate these in the living mind; in the social mind as well—indeed, this above all." He went on to conclude his lecture on a transcendent note that would have surely pleased Otlet.

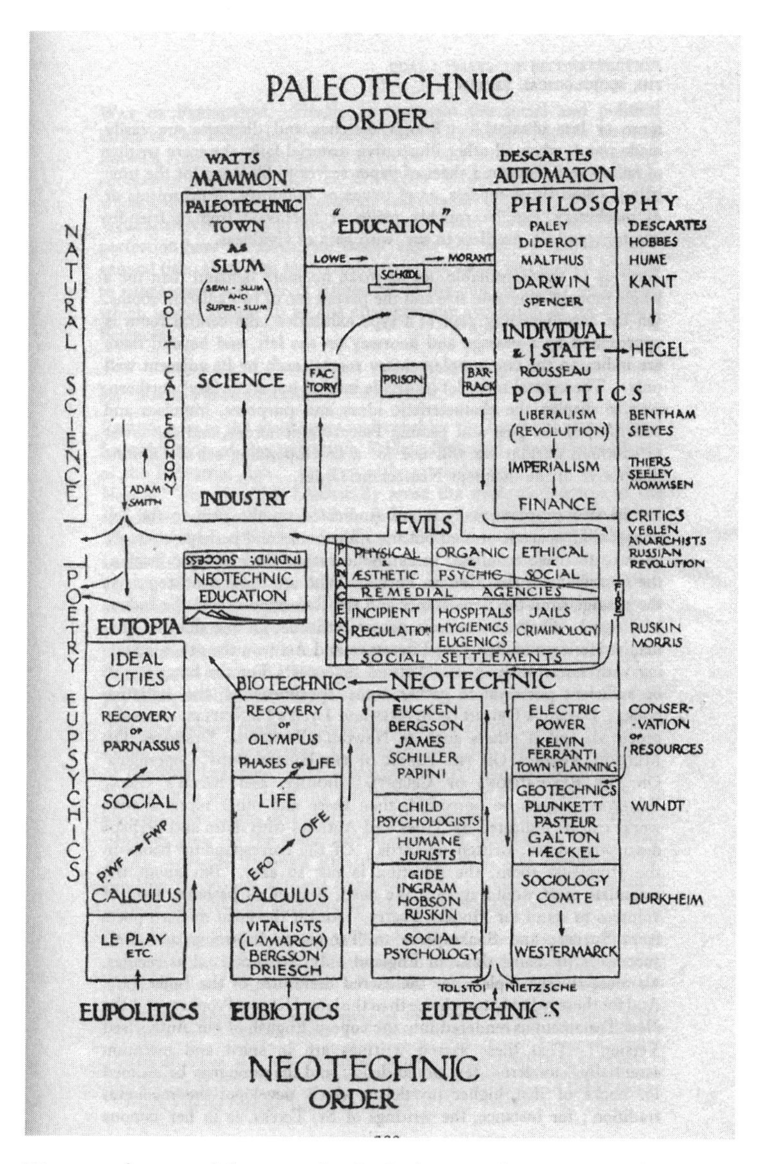

Diagram of a 1910 exhibition at the Outlook Tower, by Victor Branford. From Branford, "The Background of Survival and Tendency as Exposed in an Exhibition of Modern Ideas," *Sociological Review* 18 (1926): 207. Image courtesy of Chris Renwick.

"Beyond the attractive yet dangerous apples of the separate sciences, the Tree of Life thus comes into view."[7]

Geddes felt that the best way to shape the social mind would involve a new kind of teaching museum, designed to educate members of the public about their immediate environment and its relationship to the wider world. The Outlook Tower marked Geddes's first experiment with what he eventually came to call an Index Museum (a term he borrowed from the paleontologist Richard Owen, the man credited with inventing the term "dinosaur").[8] Geddes envisioned an institution devoted to presenting a unified overview of the intellectual world, arranged according to a master classification scheme inspired—like Otlet's Universal Decimal Classification—by positivist conceptions of a rational order to the sciences. Like Otlet, he had been strongly influenced by Auguste Comte's ideas about sociology as a tool for effecting change in the world, and of the importance of approaching the other "preliminary" disciplines—like biology, physics, chemistry, and mathematics—in the context of their relationship to human society at large. These subjects "are not so purely abstract or externally phenomenal as their students have mostly supposed," he argued, "but are each and all of them a development of the social process itself."[9]

Embracing that synthetic view of human knowledge, Geddes imagined his new museum along the lines of an encyclopedia, in which the entries would be presented not in book form but mapped on a wall and displayed "as an orderly series of labels":[10] a kind of 3-D interface, closely tied to a particular geographical location. To ensure that the museum connected with each visitor at a personal level, he grounded each exhibit in locally observed phenomena, by presenting "universal classes of things and facts by displaying locally generated or found exhibits of these classes." In this way, the Index Museum would orient each visitor to a spectrum of human knowledge

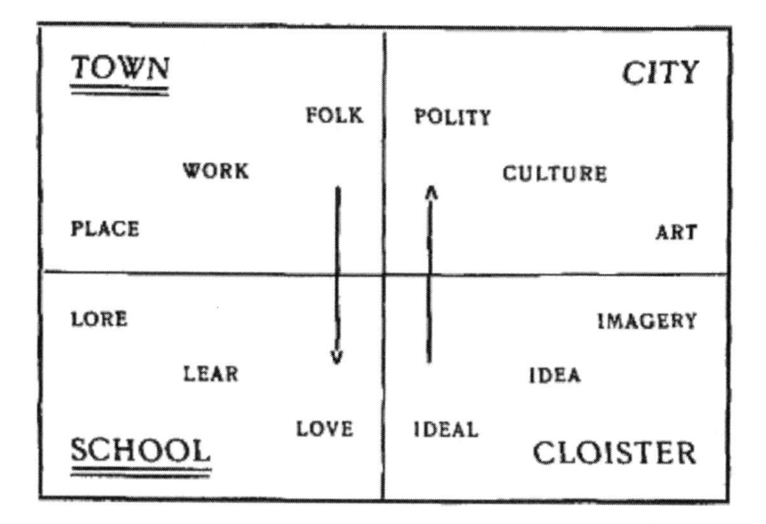

Example of one of Geddes's early "Thinking Machines." From Geddes, *Civics: As Applied Sociology* (1904), 89.

that was broad but always rooted in the immediate experience of the surrounding area.[11]

Whereas traditional museums had functioned largely as cabinets of curiosities, Geddes's museum would play a fundamentally different role. His museum would orient people to the external forces and, over time, improve their collective lot. Geddes also explored the possibilities of what he called Thinking Machines—diagrams of complex ideas that relied on visual logic rather than linear narrative forms. The American historian and *New Yorker* architectural writer Lewis Mumford called it the "art of ideological cartography."[12]

In 1910, Geddes traveled to Brussels at the invitation of his friend Paul Otlet. The two had met years earlier at the 1900 Paris Expo and afterward had exchanged occasional letters. They bonded over their shared interest in Comte's positivist philosophy, which had inspired both men in pursuit of their utopian schemes. They also shared a particular interest in the problems of classification and intellectual synthesis. Soon after their initial meeting, Geddes asked Otlet and La Fontaine for their help with creating bibliographies for the public library in Dunfermline, Scotland. The town had recently received an influx of library funds from its most famous native son, Andrew Carnegie. The three men briefly explored creating an outpost of the IIB in Scotland. Those plans failed to materialize, but their correspondence would continue for years, propelled by their shared commitment to social progress and to the knowledge engines that would drive it. Since meeting in Paris a decade before, they had pursued ambitious public-minded projects: for Otlet and La Fontaine, the Universal Bibliography and the Union of International Associations; for Geddes, the Outlook Tower, the Index Museum, and Thinking Machines. Geddes took a strong interest in the Belgians' vision of cooperation between international institutions and for a time worked with Otlet on a proposal to form an International League of Cities. Geddes's work had also inspired Otlet to begin exploring the possibilities of museums.[13]

In the wake of the 1910 Universal Exposition in Brussels, Otlet and La Fontaine had managed to exert their influence to persuade the government to grant space for a new international museum housed in one of the now empty exposition buildings, planting the seeds for the eventual Palais Mondial. A decade earlier, Geddes had led an unsuccessful campaign to convince the Parisian government to preserve the buildings from the Paris Expo. Now, he came to help them find a suitable location for the new museum. Eventually, Otlet and La Fontaine worked out an arrangement to take over part of the Palais du Cinquantenaire, originally commissioned in 1880 by King Leopold II to celebrate Belgium's fiftieth anniversary of independence.

At first, Otlet and La Fontaine populated the museum with a hodgepodge of artifacts salvaged from the Brussels Exposition, ranging from maps and printing devices to microscopes and model airplanes, as well as a large collection of administrative documents donated by the Spanish government. Ultimately, they hoped that each country would take responsibility for organizing its own exhibit to portray its social, cultural, and political life, as well as its "natural and artistic wealth." The Union would also create a series of comparative exhibits to survey "the chief facts of universal history and the various phases of civilization," as well as surveying a litany of other topics, from the grand—philosophy, science, medicine, and the law—to the highly particular—the history of the postal service,

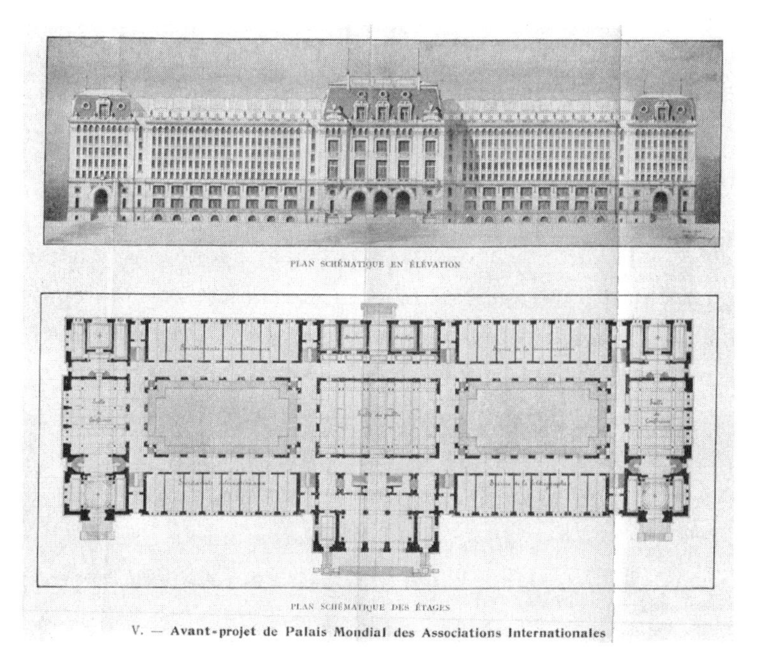

Preliminary plan for the World Palace of International Associations (*Avant-projet de Palais Mondial des Associations Internationales*). From Publication no. 25, Office Central des Associations Internationales (1912). Reproduced with permission from the Mundaneum.

interoceanic canals, submarine cables, and wireless telegraphy. Taken together, it would all add up to "a vast geographical and ethnographical museum."[14]

In hopes of accelerating their efforts and building a wider base of international support, Otlet and La Fontaine solicited funding for the Union of International Associations from the Carnegie Foundation in America, which had already demonstrated support for similar initiatives designed to promote the public good. Between 1881 and 1991, the endowment funded more than 2,500 public libraries, as well as numerous university libraries. La Fontaine wrote a letter pleading with the institution for funding to help complete the project. "Internationalism must be made conscious," he wrote in his typically fluent English,[15] outlining the plans for the Union of International Associations and its hope to build an International Museum in Brussels. The endowment looked favorably upon the application and committed funds to support it every year between 1910 until the outbreak of war in 1914.

Andrew Carnegie himself would later pay a personal visit to the International Museum. Otlet and La Fontaine arranged a formal dinner sponsored by the Union of International Associations, with Carnegie seated at the table of honor while the attendees dined on salmon and partridge. A grainy film reel preserved in the Mundaneum archives shows Carnegie touring several exhibit rooms with a coterie of men and women, including La Fontaine and a voluble Paul Otlet, leading Carnegie around while gesturing excitedly at the newly installed exhibits.[16] The trip evidently went well. "I never enjoyed a visit more," Carnegie later wrote. "The Directors of the International Centre have shown themselves clear and far sighted. The future generations will admire their combined efforts."[17]

With support from Carnegie and their ongoing government stipend, Otlet and La Fontaine continued to expand the collection.

Andrew Carnegie (center) visiting the International Museum in Brussels, with Paul Otlet (to his left) and Henri La Fontaine (to his right), 1913. Reproduced with permission from the Mundaneum.

By 1913, the museum occupied sixteen rooms in the Palais du Cinquantenaire, displaying 12,000 objects touching on a vast number of seemingly unrelated subjects: for example, the state of global agriculture, the recent explorations of the North and South Poles, the treatment of cholera, and the development of Esperanto.[18] Almost 13,000 visitors a year came to view the attraction.[19]

The structure of the museum reflected Otlet's fascination with Geddes's ideas about exhibits. The Scotsman's work offered a path beyond mere classification; it represented a new way of thinking about the spatial representations of knowledge. By incorporating visual indexing techniques and assigning labels to clusters of related documents, Geddes also imagined the museum evolving into a kind of encyclopedic enterprise. "That is we may think of it as an Encyclopaedia

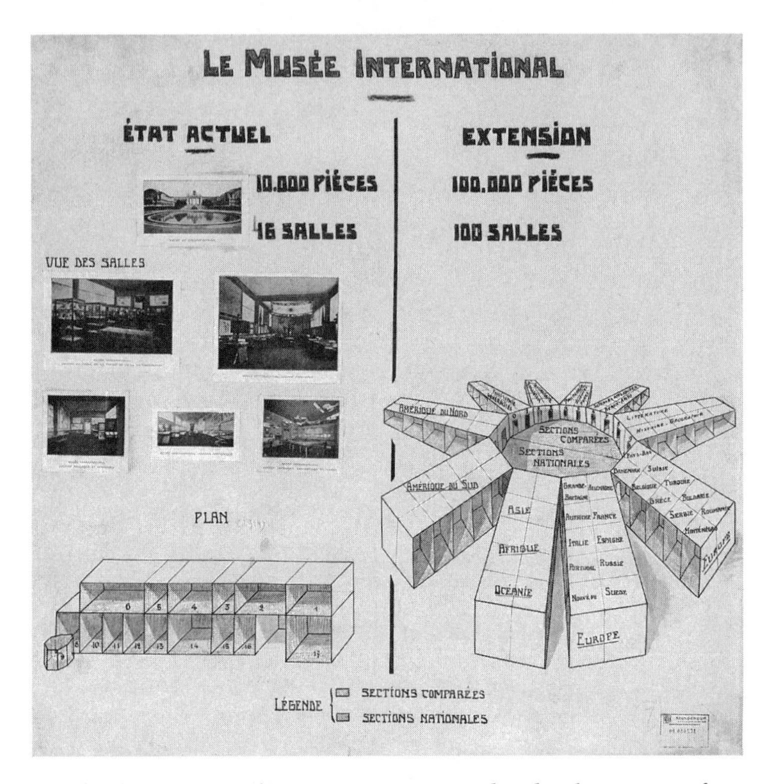

Plan for the International Museum, c. 1913. Reproduced with permission from the Mundaneum.

of which the articles may be imagined printed separately," Geddes wrote, "and with their illustration and maps condensed and displayed as an orderly series of labels; labels to which specimens are then as far as possible supplied; so that over and above the description, the image, the interpretation of the thing, you can see the thing itself in reality, if possible, or in reproduction or model as the case may be."[20]

Both Geddes and Otlet also shared a slight disdain for traditional books—no small irony given that they both published prolifically during their lifetimes. Their encyclopedic ambitions went far beyond the traditional print-bound conceptions; indeed, they saw their encyclopedia projects as transcending the conceptual and physical limitations of printed books. Although their ideas sprang from similarly idealistic visions and they shared a number of viewpoints, they also diverged in important ways. Geddes imagined his museum as what the historian Pierre Chabard calls "a spatial book"—expanding the contents of written material to fill a room. Otlet harbored a more expansive vision, in which the universal book would ultimately take the form of an entire World City, one whose contents might be read in the same way that medieval Europeans "read" the architectural features of the great cathedrals. Whereas Geddes took a "bottom-up" approach for his Outlook Towers, envisioning them as loosely federated installations, each rooted in a particular place that would broaden visitors' horizons by presenting global information from a local perspective, Otlet's

Paul Otlet (near center) with visitors to the Palais Mondial, 1930s. Reproduced with permission from the Mundaneum.

vision assumed a centralized, top-down administrative hierarchy from the outset. Whereas Geddes aspired instead "to prolong the circular horizon of each particular place,"[21] Otlet hoped for a powerful central administration to serve as "the neutral center of a pacified world."

The museum that Otlet now envisioned after his discussions with Geddes would incorporate exhibits about every country on earth, featuring materials depicting each country's social and political systems, natural resources, economic activity, culture and artistic output, as well as its contribution to world culture. The museum would also include broader-themed exhibits intended to portray "all that is general, universal and essentially human," including an overview of the sciences, philosophy, and—with a nod to his positivist outlook—"the various phases of civilization."[22]

The Outlook Tower was an inspiration for what his International Museum might become. Although the two men shared similar hopes of social transformation, Otlet pursued a more explicitly universalist approach. Geddes had, as we've seen, encouraged his visitors to engage with his museum in an inductive way—starting with the highly local, particular situation of Edinburgh, then working their way out (down) into progressive levels of abstraction. Otlet took precisely the opposite approach, grounding his museum at an explicitly "universal" level of understanding. His basis was not the streets of Edinburgh but what he called the "Sphere of the World," a symbolic representation of the unity of all knowledge arranged in three interconnected organizing principles: Objects, Space, and Time (echoing Geddes's notion of work, place, and folk). As visitors entered the main hall of the museum they would find three groups of chambers representing each of these principles.

The International Museum gathered momentum quickly, consuming more and more of Otlet's time as he sought to build up the collection that he saw as a natural extension of the Universal Bibliography, one that simply happened to exist in three dimensions. In the years to come, Otlet would push that vision even further: extending the boundaries of the unified world he imagined to include the design of cities, the governance of nations, and beyond.

6

Castles in the Air

Hendrik Christian Andersen was still an infant in 1873 when his father, Anders, ran off to America. A onetime reindeer herder, the elder Andersen had as a young man moved to Bergen, Norway, where he met and married a woman named Helene and fathered two children. For a time the young family seemed happy enough. But by the time he abandoned them, Anders had stopped working regularly, often disappearing for days on end, only to be brought home by his friends, too drunk to stand. One day, he made the acquaintance of an American painter who was looking for a man to bring two miniature Norwegian horses back to Newport, Rhode Island. Lured by the prospect of adventure, he departed for the New World, leaving his wife and two boys behind. Penniless, Helene sold her furniture to pay off her husband's debts, then with the boys in tow crossed the fjords to move in with relatives, toting her only possession: a trunk of house linens. She tried to support the family doing odd jobs until eventually she received a letter from her husband, along with $100 and an invitation to join him in Newport.

Helene and her two little boys had a miserable three-week ocean crossing. The ship's captain cut the engines at one point to limit the damage from a seven-day storm, setting the ship adrift amid the massive ocean swells. The food ran out, and the Norwegian crew briefly mutinied. When the ship finally reached the port of

New York, it disgorged a group of sickly passengers struggling to get their bearings. Anders Andersen failed to meet his family at the port, but Helene and children eventually made their way to Newport and finally reunited. Alas, not much had changed: He still preferred drinking to work. The family spent their first few weeks in a basement with no furniture and barely anything to eat. Helene eventually found work as a maid to support the family, and as soon as the boys, Hendrik and Andreas, were old enough they too went to work—shining shoes, selling papers, and running errands for neighbors.

Young Hendrik worked for a time driving a grocery wagon, with which he would occasionally make deliveries to the "cottages" of Gilded Age Newport, giving him his first glimpse of Gilded Age opulence. Each summer assorted Astors and Vanderbilts would arrive for the summer season, engaging in a whirlwind of society dinners and dances in the evening, tennis and polo in the daytime. While the wealthy summer residents lived behind an aura of privilege symbolized by high garden walls, in truth they enjoyed little privacy; word of their goings-on circulated freely around town, by way of the network of household servants and tradespeople who kept their households running, including the Andersens.

After leaving school at thirteen, Hendrik worked briefly as a telegraph operator at the Ocean House, where Wall Street titans like Pierpont Morgan would go to receive the latest stock quotes. He later found work at the Gardner House, located across the bay in Jamestown, sweeping the wooden piazzas where the debutantes plied their charms and busing dishes from the dining room that hosted elegant society dinners. Feeling homesick, one night he borrowed Mr. Gardner's boat and sailed across the bay to see his family. Enraged, Gardner promptly fired him the next day. Back in Newport, Hendrik managed to find new work as an assistant carpenter and ship-painter, discovering an aptitude for crafting material with his hands. At night he took classes from a teacher named Miss Ellery, who held something equivalent to an adult-education class in her home at night. Artistic ability seemed to run in the family; his older brother, Andreas, had made his way to the Cowles Art School in Boston a few years earlier and invited Hendrik to follow him there. Hendrik took his brother's advice and pursued his art studies with enthusiasm. A few years later, he followed his brother to Europe, continuing his studies in Paris at the Académie Julien and the École des Beaux Arts. In 1896, he made his way to the city that he would eventually come to call home: Rome.[1]

In 1899, Hendrik's friend John Elliott brought a guest to his basement studio at 53 Via Margutta in Rome: the American writer Henry James. Although separated in age by more than three decades, the two men struck up a friendship—the famous older writer and the handsome, idealistic young sculptor. James seems to have developed an instant liking for Andersen, dubbing him "that most loveable youth." At one point he wrote to his brother William to tell him about his new acquaintance, this "sincere and intelligent being." Andersen in turn described his relationship with James as one of "brotherly love which with the passing of years grew ever deeper as the invisible vails [*sic*] of suffering and sorrow united our spirits with the tenderest affection."[2] Over the years that followed, the two men exchanged more than seventy-five letters, whose contents reveal a thoughtful if occasionally tense relationship.

James scholars have noted that the writer's fondness for Andersen seemed to carry more than a hint of homoerotic attraction. He often ended his letters with effusive codas, such as "I pat you on the back, tenderly, tenderly"; "Ti abbraccio bene, Enrico caro, ti stringo caramente" ("I embrace you well, dear Henry, I hold you dearly"); "I feel, my dear boy, my arm around you. I feel the pulsation, thereby, as it were, of your excellent future and your admirable endowment." Andersen seems to have reciprocated the affection, sometimes calling himself to James "your son Hendrik" or, at another time, fearing that James might, like an angry father, "lay him across his stout knee and spank him on both cheeks of his fat backsides."[3] It seems unlikely that anything happened between them, but there is little question that James harbored deep feelings toward Andersen, feelings that often found expression in descriptions of physical affection.

The relationship between James and Andersen forms a central chapter in *The Master*, Colm Tóibín's novelistic portrait of James that explores the writer's obsession with this "golden youth" who was possessed of such "good lucks and unsettling charm."[4] "Unsettling charm" seems an apt description of Andersen's persona. He was determined to secure fame for himself and was more than willing to seek out famous men in pursuit of his own ambition. Tóibín points out the parallels between Andersen and a character that James had conjured early in his career, in one of his minor novels, *Roderick Hudson*: an ambitious but rough-around-the-edges young American sculptor who tries to make a name for himself in Rome.[5] "Both of them were doted on by a worried mother back at home," writes Tóibín, "and both, once installed, were watched over by an older man, a lone visitor, who appreciated beauty…and kept passion firmly in check."[6]

Hendrik's sister-in-law Olivia Cushing Andersen also noticed the intensity of their relationship but saw nothing particularly wrong with the idea of an older man taking an interest in a talented young male artist. "There is with certain intellectual men a special charm and attraction about an intellectual youth," she wrote, "which they seem not to find in their contemporaries or in the opposite sex." As to the question of whether such an attraction might have a physical component, she avoided a direct answer, writing in her diary that the "physical and the mental are so... interwoven that it becomes difficult to say."[7]

Andersen, for his part, felt that James possessed special insight into his life. They would engage in long discussions in the summer nights, talking over their respective pasts, and the places they had in common: Newport, Boston, Paris, and Rome. "He alone knew the bitterness and hardships that had crushed my spirit from early childhood, the confessions of which he Treasured and held sacred," Andersen recalled. "These confessions drew him nearer and dearer to me... we both longed for each other."

Despite his affection for the young man, James also seems to have entertained occasional doubts about his new friend's character. Much as he seemed to enjoy Andersen's company, James never introduced him to his literary circle. Perhaps with reason. Edith Wharton's friend, the art patron and collector Isabella Stewart Gardner, referred at one point to his sculpture as "execrable." But at times James seems to have been put off by his friend's disposition, at one point writing to his brother William that Andersen was "handicapped by a strange, 'self-made' illiteracy and ignorance of many things."[8]

Whatever his doubts about Andersen's character, James seems to have genuinely admired Andersen's sculpture. He was "lavish almost to the point of exaggeration in his praise," as Andersen later put it. But he always preferred Andersen's smaller portrait busts to the operatic, oversized figures that characterized much of his later work. At one point Andersen created a bust of James that shows the writer striking a pensive pose, but he maintained that he had little interest in portraiture, recalling that James had found it "painful to realize" "that I cared so little for this appealing form of art." Andersen spent most of his time working on his grand, often grandiose statues. His sculpture evinces an affinity for the classical tradition that enveloped him in Rome, with its heroic figures—usually rendered naked—seemingly striving to represent some noble ideal. "His work was a set of large, energetic gestures," writes Tóibín, "commensurate with his dreams." Those dreams were large indeed.[9] Over the years that followed, Andersen began to extend his vision beyond the realm of individual sculptural works, ultimately hatching a grandiose plan to make his oversized sculptures the centerpiece of an urban center called the World City.

In 1902, Hendrik received news that Andreas had died of tuberculosis, leaving behind his widow of one month, the heiress Olivia Cushing Andersen. Stricken with grief, the young woman turned for comfort to her brother-in-law, traveling to visit him in Rome. Joined in grief, the two soon grew into close companions and would remain close for the rest of their lives. Their relationship, however, never turned romantic. Andersen once told his sister-in-law that "the meeting of bodies is nothing and never can be. It is the meeting of mental forces that count[s]."[10] As Hendrik and Olivia spent more and more time together, they seem to have developed a close mental and even spiritual bond. Olivia would spend much of her inheritance promoting Hendrik's career; her diaries depict a woman wholly devoted to the talented young man she came to call "my brother."

With Olivia's help, Hendrik began fleshing out these plans. In 1907 he made contact with a prominent Parisian architect, Ernest Hébrard, to create a series of architectural drawings, and slowly the two began work on describing their vision in the form of a massive two-volume book. At one point Hendrik referred to the project as their "baby." "So the baby we have made is cutting its teeth and perhaps will grow up to speak in many languages.... I am sure its eyes will always be deep and blue, its hair rich and golden, and there will be something round and full of sweet dimples about it that all nations will kiss."[11]

Under Andersen's direction, Hébrard began work on developing the architectural drawings for the World City.[12] They imagined it as a three-mile-long rectangular settlement, three miles long and a half-mile wide, featuring a downtown district marked by broad avenues and monumental edifices. It would include administrative buildings, a bank, and various "temples" (*palais*) devoted to the pursuit of art, music, drama, and other cultural endeavors. There would be a world zoo, botanical gardens, and a sports center designed to host the Olympics, featuring a large stadium and colossal swimming pool (or "natatorium"). The area between the stadium and the Grand Canal would feature a recreation area, with a ball club, skating club, tennis club, and kindergarten. At the base of a grand avenue lined by the palaces of nations there would be a great circle ringed with the institutions of the new world government: an international court; ministries of industry, agriculture, medicine, and science; as well as the great library and a palace of religions. They all fanned out from the great circle, forming a kind of lopsided mandala around a central Eiffel Tower–like Tower of Progress, a symbolic center for scientific research that doubled as a radio tower to send and receive wireless signals from all over the world.

This new city was designed to support a new form of social interaction, expressed through architecture, evoking the idealism and over-the-top spectacle that was evident just a few years earlier at the 1900 Paris Expo. Indeed, scholar Wouter Van Acker has argued that the "monumentalness" of the Expo directly influenced the tone of the World City designs, "in order to evoke not the power of the nation but the grandeur of a possible world state."[13] Andersen took particular inspiration from his surroundings in Rome, the fallen capital of the ancient world. The city "touched a deeper chord in the heart and conscience of man," he wrote, "and brought him to a full sense of the divine meaning and value of his soul."[14]

Drawing of the World City by Ernest Hébrard, from Andersen and Hébrard, *Creation of a World Centre of Communication*, 1913.

This is perhaps why the World City would feature a Fountain of Life, a large round basin with a blue mosaic of the globe at the bottom, surrounded by Andersen's statues depicting four groups of people with children, representing the four seasons as well as the "shades of sentiment that bind humanity together." Symbolizing the flow of pure energy traversing the city, the fountain would stand for the reservoir of all knowledge, "fed by the whole world of human endeavor in art, science, religion, commerce, industry and law."[15]

On a trip back to America in 1911, Hendrik and Olivia Andersen met a gentleman named Urbain Ledoux, who was working at the time for the Boston branch of the Peace Society. A former U.S. consul in Prague and sometime Harvard lecturer, he was evidently a natural diplomat and "a great collector of facts," as Olivia would later observe in her diary. Ledoux impressed them right away with his enthusiasm and apparent passion for the internationalist cause. He also seemed to have the right connections in the world of international politics and finance. He had served as a delegate to the first International Peace Conference in the Hague in 1899, and later worked with Andrew Carnegie as a member of his international peace committee. Hendrik and Olivia felt certain that such a well-connected American could help promote their cause. "Mr. L. seems to be exactly the man needed to bring forward our plan."[16] The Andersens hired him for $5,000 plus $1,000 travel costs to bring his family over to Europe for one year, hoping he could bring his diplomatic skills to bear in helping them sell their vision of a World City to European governments and international institutions.

Soon after they first made contact, Ledoux suggested that Andersen should consider sharing his plans for the World City with Paul Otlet and Henri La Fontaine, whose work he had come to know through his involvement with the international peace movement—a community in which La Fontaine had by then earned a prominent reputation, having served as president of the International Peace Bureau since 1907. Ledoux recognized Andersen, Otlet, and La Fontaine as kindred spirits. "Both plans have frequently been considered as absolutely Utopian," he wrote in a letter to Andersen that year, "and their creators expert dreamers of the impossible." "My faith is pledged in both." Ledoux knew of Carnegie's support for the Belgians' projects, and he suggested to Andersen that if he could hitch his project to theirs in the spirit of international cooperation, they might together convince Carnegie to fund the World City as part of an integrated project with the Belgian plan for a unified network of governments and international associations. "There is no doubt that as Mr. Carnegie gains better knowledge of this special work—in which I found him keenly interested—and should your project be in some way linked to it... that he might be interested in endowing the cause."[17]

Ledoux made plans to meet with Otlet and La Fontaine in Brussels, inviting Andersen's architect, Hébrard, to accompany him for a visit in September 1911. Afterward, he wrote back excitedly to tell Andersen that Otlet had met them "with enthusiasm," showing off a potential site near Brussels that they hoped the government might consider making available for the World City project. The parcel of land included 4,000 acres of forestland near Tervuren, with an already cleared park of 280 acres—plenty of room for the city plan that Andersen and Hébrard had devised. Ledoux suggested that Otlet and La Fontaine also had a notion of how to pay for the project, mentioning that they had "a keen eye on the Leopold millions left in trust for the benefit of mankind,"[18] a trust funded almost entirely with his personal spoils from the Congo.

As Ledoux pressed ahead in forging a partnership with the Belgians, Hendrik and Olivia Andersen were beginning to entertain some doubts about their new American associate. "I have not been able to prevent a feeling of mistrust of that man Ledoux," wrote Olivia in her diary in September 1911. Those doubts continued to grow in the weeks ahead. "Ledoux wants an easy fame," she wrote two months later, "to reap what he has not sown, and talk a great deal about the greatness of the achievement." And while Olivia had kept Hendrik's dreams afloat with the regular allowance from her inheritance, she nonetheless felt surprisingly indignant at the prospect that the World City could be financed by an inheritance from the late Belgian king. "Somehow I feel like vomiting," she wrote, "and rejecting all suggestions of trust funds and millions of this or that man. I seem to hear the ravens gathering and hear the squawk of their prayers."[19]

Nonetheless, they continued to move forward with the collaboration. In November 1911, Ledoux wrote excitedly to his employer: "Very favorable news comes from Brussels and I believe they are creating interests in all directions for the realization of the project."[20] Andersen's collaborators urged him to travel to Brussels as soon as possible. Back in Rome, Olivia recounted discussing Ledoux's letter with her brother-in-law. "Those gentlemen," Andersen said indignantly, "can come and see me."[21]

Andersen eventually softened his stance, however, and struck up his own correspondence with Otlet and La Fontaine, sharing the drafts of his plans and soliciting their input. After a visit to Hébrard's offices in Paris in February 1912, Otlet and La Fontaine wrote to Andersen to tell him that they found his plan "grand in its conception," going on to say that "the center you have dreamed of in its architectural realization, we have imagined in its functional activity."[22] Andersen and Otlet continued to correspond regularly in the months ahead, exchanging plans and exploring the possibility of collaboration.

Despite their great professions of mutual admiration, however, Andersen's relationship with Otlet and La Fontaine would soon turn contentious, as the men struggled to define the terms of their working relationship. For all of Ledoux's excited talk of Carnegie's endowment and King Leopold II's millions, Andersen soon discovered that their potential collaborators were perpetually cash-strapped. Nonetheless, Ledoux and Hébrard pressed forward in their discussions with Otlet and La Fontaine, developing a proposed plan to merge the World City project with the Central Office in Brussels, to affiliate their project with the 132 international associations already involved with the Union of International Associations.

On March 12, 1912, Andersen sent a letter to Otlet and La Fontaine, affirming his interest in their work. "I am becoming more and more familiar with the very noble work that you both have been pursuing," he wrote, "and the more I become familiar with it, the closer my sympathies are drawn toward the high human ideals that you have consecrated your life to accomplishing." He went on to acknowledge the internationalist viewpoint that had informed their work. "I now realize that my dream of harmony among all human endeavors coincides with the great and righteous movement of the age, that is, internationalism."[23]

They put together a detailed plan for the collaboration, including a shared effort to promote Andersen's book and arrange public exhibitions of the architectural plans. These protestations of support notwithstanding, Andersen retained his own reservations about the project, and about his man Ledoux. Evidently frustrated that his accomplices were beginning to overstep their responsibilities, Andersen began to assert himself. On July 10, 1912, after learning that Otlet and La Fontaine had made an initial overture to Andrew Carnegie regarding the World City project, he wrote to tell the men he was "indignant" at their using his name in this fashion. He also inquired as to whether they had heard from Mr. Ledoux, since it had now been three weeks since he last heard from his employee.

Otlet replied with a reassuring letter, offering to discuss the matter in person—and telling him that they had not seen Ledoux since he left for Paris several days beforehand.

Frustrated with Ledoux, Andersen took the reins himself, finally traveling to Brussels to meet with Otlet and La Fontaine. Evidently the tensions had cooled, for he came away feeling optimistic about the prospects for their collaboration. Describing the meeting in a letter to Olivia and his mother on the stationery of the Hotel Metropole in Brussels, he wrote: "This morning I saw Senator La Fontaine & M. Otlet—they have been with me until a short time ago, and I am sure it is a conference that they will both remember for some time. Eight-hours of almost uninterrupted talk." He went on to describe his impression of both men as "charming and deeply interested; they are fully of sympathy and very kind." Despite that agreeable first impression, the visit seemed to raise questions about the viability of their project. "The whole international organization is based upon scientific lines, and well organized," he wrote, but "how much of humanity will fit into the wheels of their caravan and how far the caravan will move toward a happy solution of human desires, is a little uncertain in my mind." Andersen seemed concerned that their system was too regimented and tightly bound to encompass the entirety of humanity. "I am afraid their lines are in many places too narrow and their rules too material to meet a grand world movement."[24] He resolved to keep his reservations to himself, however, and moved forward with the partnership.

Later that month, Andersen finally resolved to let go of Ledoux. "I dread kicking Ledoux out of my service; yet I do it with pleasure and decision," he wrote. "I dread it for his wife and family's sake." A few days later: "Ledoux came to dinner drunk. Got rid of him, talks through his hat. Taken in by him." On August 28, Ledoux failed to show up for a planned meeting at 4 p.m. By 5:30, Andersen had

had enough. "No Ladeux [*sic*], foolish fat stupid man." Finally, Andersen severed his ties with Ledoux once and for all.[25]

Years later, Ledoux would resurface in New York City, having adopted a new name: Mr. Zero. After converting to Buddhism, he opened a shelter in the Bowery named The Tub, where more than 1,000 men per night would sleep in steamer chairs and enjoy a 5-cent dinner of soup and bread while listening to lectures by visiting speakers, who paid $50 for the privilege of addressing the audience of "old bucks and lame ducks."[26] Perhaps the idealism and self-promotional energies of Andersen and Otlet had rubbed off on him. A regular fixture in the New York daily newspapers of the 1920s, he led a procession of homeless men down Fifth Avenue in the Easter Parade, and once gained an audience with Warren Harding while wearing sackcloth and ashes, after threatening to camp out on the steps of the White House.[27]

With Ledoux out of the picture, Andersen, Otlet, and La Fontaine continued to correspond about their plans. Otlet urged Andersen to participate in an upcoming exhibit on urban planning that Patrick Geddes was organizing in the Belgian city of Ghent. Andersen offered to send copies of his plans but balked at the prospect of sending a full-fledged exhibition, concerned at the cost of transportation and the need for space to display more than fifty designs, many of which stretched ten to fifteen feet in length. But the exposition could offer no financial support; Andersen would have to arrange to transport and display the designs himself. Hendrik and Olivia began to suspect the destitution of the men whom they had once thought might steer them to the capital needed to build the World City.

The Ghent exhibition went forward without Andersen, but he nonetheless agreed to join forces with the International Union of Associations. The Union would lend its backing to the World City project, which, in turn, would commit to make the Union—and Otlet's pet projects, including the Universal Bibliography—an integral component of the proposed World City. With his book ready to publish and bolstered by the backing of the international associations, Andersen experienced a renewed burst of optimism. A few days after writing to Otlet, he picked up his pen to write to Henry James, with whom he had yet to share any of his plans for the World City. Andersen sent James a copy of his book, asking whether his friend would consider lending his literary talents to the cause.

"My dear Henry James," he wrote. "The work I am up to my eyes in is almost finished, that is the planning of an 'International WORLD CENTER' of communication. These plans I have been at work upon some twelve years. I have never spoken of them, little by little as this City grew in my mind." He went on to mention Otlet and La Fontaine, whom he referred to as the "captains of an International Central Office" in Brussels. "I am in touch with the leading men of this office," he wrote. "They have seen my work in Paris and approve of it." He then beseeched James to exert his influence in support of the project and provide feedback on his manuscript. "Will you give me a tip?" he asked. "Will you read it or have it read to you before I send it to the press?"[28]

Weeks passed with no response from James. Finally, evidently unable to contain his apprehension, Andersen sent a follow-up note, pleading with his friend for a response. "I have not received any reply to my last letter and I feel like a young lady about to give birth to her first baby," he wrote. "I confess I am uneasy and need assurance."[29] Little did he know that his letter would cross paths with James's reply in the mail. Here he encountered few of the usual expressions of affection and support that he had come to expect from James over the years. Instead, the writer responded with a withering rebuke.

"Brace yourself," James wrote:

> Your mania for the colossal, the swelling & the huge, the monotonously & repeatedly huge, breaks the heart of me for you.... The idea, my dear old Friend, fills me with mere pitying Dismay, the unutterable Waste of it all makes me retire into my room & lock the door to howl! Think of me as doing so, as howling for hours on end, & as not coming out till I hear from you that you have just gone straight out on to the Ripetta & chucked the total mass of your Paraphernalia, planned to that end, bravely over the parapet & well into the Tiber. As if, beloved boy, any use on all the mad earth can be found for a ready-made city, made-while-one-waits, as they say, & which is the more preposterous & the more delirious, the more elaborate & the more "complete" & the more magnificent you have made it. Cities are living organisms, they grow from within & by experience & piece by piece; they are not bought all hanging together, in any inspiring studio anywhere whatsoever, and to attempt to plank one down on its area prepared, as even just merely projected, for us is to—well, it's to go forth into the deadly Desert & talk to the winds.[30]

In a diary entry of August 31, 1913, Olivia described how she and her brother-in-law had taken James's rejection. "Henry James has lifted his hands, and shaken his head, pouring forth torrents of remonstrance. He would do for a grotesque figure in a formal garden, with a cascade rushing out of his mouth. Nevertheless, one can but try; bend the knee before these people while there is any hope of approval.... No matter how much one need humble oneself to obtain it and no matter how many rebuffs one receives!" Crestfallen, Andersen nonetheless resolved to press his case. That same day, Andersen wrote again to James, hoping to convince him to reconsider his opposition to the project. "I cannot believe, my dear good friend, that you have any cause to be either disgusted or sorry for me."[31]

James refused to reconsider, but in his response of November 28, 1912, he tempered his criticism a bit, thanking Hendrik for sharing his "fond and devoted dream, in which you are spending your life, as some Prince in a fairy tale might spend *his* if he had been locked up in a boundless palace by some perverse wizard, and, shut out thus from the world and its realities and implications." Nonetheless the World City was still in his view a "vast puzzle," putting him in mind of "colossal aggregations of the multiplied and the continuous and the piled up, as brilliant castles in the air."

James continued to show no interest in promoting the project, but Andersen seemed unable to take a hint—he pressed James once more for his support. Finally, on September 4, 1913, James wrote the letter that would end the dialogue once and for all:

> Do you think, dearest Hendrik, I <u>like</u> telling you that I don't, & can't possibly, go with you, that I don't, & can't possibly, understand, congratulate you on, or enter into, projects & plans so vast & vague & meaning to me simply nothing whatever? . . . I don't so much as <u>understand</u> your very terms of "World" this & "World" the other & can neither think myself, nor <u>want</u> to think, any such vain & false, & presumptuous, any such idle & deplorable & delirious connections. . . . The World is a prodigious & portentous & immeasurable affair, & I can't for a moment pretend to sit in my little corner here & "sympathise" with proposals for dealing with it. It is so far vaster in complexity than you or me, or than anything we can pretend without the imputation of absurdity & insanity to do it, that I content myself, & inevitably <u>must</u> (so far as I can do anything at all, now,) with living in the realities of things, with "cultivating my garden" (morally & intellectually speaking,)

> & with referring my questions to a Conscience (my own poor little personal,) less inconceivable than that of the globe.[32]

This time Andersen seems to have gotten the message. Olivia described her reaction to James's latest dismissal. "An answer came tonight from Henry James," she wrote. "I must copy it, it is worth keeping. If one were depending upon him for counsel or advice! What would grow in the world if he had his say? Reality! Reality? His 'reality' is the perishable, the fading." While praising James's intellect and stature, Andersen's devoted sister-in-law went on to excoriate what she saw as his small-mindedness when confronted with Hendrik's expansive vision of the future. Undeterred, she vowed to continue to help her brother-in-law press on in the noble cause that had already become their life's work. For Olivia Cushing, an aspiring writer in her own right, the sting of James's rejection seems to have struck particularly deep. "Well, it is no use expecting that minds diametrically opposed once they have met will ever cross paths again," she wrote. "The one will bury himself deeper and deeper in detail; the other will swing out further and further into immeasurable conceptions." She went on to write that while there were those like James who might maintain that all they could do is deal with their own "little private" worlds, there were "men and women, no longer so troubled by their little private ones that they cannot sink some or all of their personality into the godlike personality which longs to become manifest in us."[33]

James's rebuke marked a turning point in his relationship with Andersen. Although they exchanged a handful of additional letters, the two men never met again. James died from a stroke on February 28, 1916.

While Andersen's friendship with James wound to a close, he continued to enjoy the enthusiastic support of La Fontaine and Otlet.

They accelerated their dialogue, exchanging views about their respective schemes, and started to explore the possibility of working together even more closely. Although Olivia harbored some private reservations about Otlet and La Fontaine's museum project—"It collects a multitude of international facts, and gathers them together in almost an infantile way," she wrote in her diary on November 23, 1913—still she seemed to appreciate the energy and enthusiasm underlying their efforts. "Nevertheless, this museum is Sen. La Fontaine's and M. Otlet's pet child, the apple of their eye, their ewe-lamb and I could but look at it with sympathy and think it would be well placed in one of the many buildings of our city."[34]

In Otlet and La Fontaine the Andersens seem to have recognized fellow travelers—men possessed of bold utopian ambitions and willing to act upon them. Hendrik wrote excitedly to Otlet about the important contacts he was making in Paris, evidently trying to assure his new Belgian correspondent of his seriousness. Otlet, in return, told Andersen more about his plans for expanding the international museum. Olivia took a particular liking to La Fontaine, describing him as "delightful, simple as a child, very intelligent, sympathetic entirely uncalculating personality and devoted to his work."[35]

Andersen continued to promote his World City project relentlessly, sending free copies of his book to Woodrow Wilson and all the heads of state of the European countries, as well as to various diplomats, business leaders, and libraries all over the world. He even managed to secure an audience with Pope Benedict. But not everyone welcomed him. The American minister to the Hague turned down his request to arrange a meeting with the Queen of the Netherlands. He also tried in vain to secure a meeting with the King of Sweden. But his connections in Belgium continued to bear fruit. With the support of Otlet, La Fontaine, and an American diplomat named Theodore Marburg, he requested an audience with King Albert I of Belgium.

Marburg assured Hendrik that his "well-known interest in the international movement and personal standing would quite justify his taking the necessary steps for an audience with the King," according to Olivia's account of their conversation. "Although he could not affirm that this would be granted, he said he would do what he could." And so he did. In November 1913, Hendrik and Olivia traveled to Brussels, where Hendrik made his way to the Royal Palace for his audience with King Albert.

"The visit to the King was even more satisfactory than Hendrik had expected," Olivia wrote the next day. "His majesty though shy is intelligent, and clearly his desire was to show his sympathy both with the idea and its originator." After conversing with the king and sharing his plans for the World City, Hendrik acknowledged that some people might view his project as little more than a dream.

"There must be those who dream for the others," said the king, "but this plan is practical, there is nothing here that cannot be turned into a reality." He offered Andersen his wholehearted encouragement. "One day it must exist."[36]

7

Hope, Lost and Found

Just as Otlet and La Fontaine's dreams were starting to bear fruit—with the World City project coming into focus, the Union of International Associations boasting more than 150 member associations, and the Universal Bibliography now surpassing 10 million entries—Europe fell apart.

The halcyon days of the belle epoque—the era of Art Nouveau, symbolist poetry, and the optimistic spirit of the 1900 Paris Expo—had given way to a rising tide of nationalism and industrial militarization, as the Great Powers jostled for influence both at home on the European continent and abroad, especially in the increasingly lucrative African colonies. Amid these growing tensions, a tangle of international treaties and alliances ensured an unsteady peace. But in June 1914, when a Serbian nationalist assassinated the Austro-Hungarian Archduke Ferdinand, those alliances triggered a domino sequence of events that soon plunged the continent into all-out war.

Just one year earlier, the Nobel Committee had met in Stockholm to announce its selection of that year's Nobel Peace Prize winner: Henri La Fontaine. Recognizing his work as president of the International Peace Bureau, his efforts with the International Union of Associations, as well as "his great documentary work," the committee chose to bestow its highest honor on the Belgian senator.

"Henri La Fontaine is the true leader of the popular peace movement in Europe," said committee secretary Ragnvald Moe in his remarks. "There is no one who has contributed more to the organization of peaceful internationalism."[1] The prize cemented La Fontaine's reputation as a world political figure, and the prize money of 50,000 Swedish kronor—worth over $3 million today—had seemed to guarantee the future of his and Otlet's projects for years to come.

Now, millions of French and German troops were massing across Belgium's borders, while the tiny Belgian army gathered horses and cannons in the Parc Cinquantenaire, surrounding Otlet and La Fontaine's museum.[2] The Germans calculated that Britain would choose not to honor its commitment to guarantee Belgium's neutrality, an agreement entered into long ago at the 1839 Treaty of London that secured the country's independence from the Netherlands. German chancellor Bethmann Hollweg famously predicted that Britain would never go to war over a mere "scrap of paper."[3] His assessment proved disastrously wrong. Britain declared war on August 4, 1914, the same day that German troops rolled into Belgium.

La Fontaine now looked on helplessly as his country plunged into war. "From the bottom of my heart," he wrote, "I am sickened by humanity's villainy and irreducible stupidity."[4]

For the first time in its brief history, Belgium found itself at war. Germany launched a vicious campaign of total war that came to be known as the "Rape of Belgium." Across the English Channel, Britons started to receive a steady stream of wire stories chronicling the mass killing, rape, and torture of civilians, destruction of homes, burning of villages, and other atrocities visited upon the Belgians. While the British wartime propaganda machine clearly exaggerated and in some cases even fabricated these stories (partly in hopes of building public support in America to enter the war) there is no doubt that a great many real atrocities took place.[5] In the town of Leuven—where Otlet had once attended school—the Germans poured kerosene on the university's 300,000-volume library (including a vast collection of priceless medieval manuscripts) then set fire to the building. They went on to raze the town, shooting numerous civilians in the streets before expelling thousands of residents at gunpoint.[6]

While the British public responded with an outpouring of compassion for the Belgians' plight, some could not help but notice the apparent karmic dimension of their suffering, given the kingdom's by then well-known history of atrocities in the Congo. A former British diplomat named Richard Casement, who had once led a British government investigation into the Congo situation, wrote in his diary upon learning of the Belgians' predicament: "I must confess, when the present 'agony of Belgium' confronts me—and it cannot well be minimized, it is in truth a national agony—I feel that there may be in this awful lesson to the Belgian people a repayment."[7]

The German occupying army of more than half a million men faced a Belgian defensive force barely a fifth that size. Its ranks included both of Otlet's recently enlisted sons. His elder son, Marcel, was taken prisoner in September at the Battle of Antwerp. One month later, Jean went missing at the Battle of Yser; his whereabouts would remain unknown for most of the war. Distraught, Otlet spent the next few years casting around for news of his lost son. Ever the tireless researcher, he spent countless hours writing letters to the Wounded Allies Relief Committee in London, the International Red Cross in Switzerland, the Swiss Catholic Mission, the Belgian government, the Spanish government, even the German government—anywhere he could think to ask for news of Jean. Each time, his search came up empty.[8]

Soon after the Germans began their occupation, Otlet published *The End of War,* a short manifesto in which he pleaded for a new world order that might lead to a general peace. While he invoked his standing as the founder of the International Union of Associations and took sole responsibility for its contents, he had the book published in the Hague, presumably hoping to avoid scrutiny by the Germans while distributing it through his own social and professional contacts. The book opens with a plea for European diplomats to put their people's well-being ahead of their own narrow interests. "Where can we find an end to the causes of war? The people hope for this outcome. They are tired, or they soon will be. They do not understand all the diplomatic squabbling. . . . What they see with clarity is evidence that the political machine is broken, that it does not respond to their sacrifices, and that there must be a new engine of governance."[9] Otlet went on to describe how just such a "new engine" might work. He envisioned it taking shape as an international organization, of course, founded on a charter to secure human rights and foster cooperation between nations. The organization that he described sounds very much like the United Nations, and the vision he described played a role in the formation of the League of Nations in the aftermath of the war.

Over the preceding quarter century, he argued, the world had seen a flurry of rhetorical activity about improving international relations at conferences, universities, and diplomatic gatherings. While a cynic might have seen these efforts as exercises in empty bureaucratic rhetoric that had ultimately failed to forestall war, Otlet saw this emerging global dialogue as the basis of a new world peace movement, and a cause for hope. "Never before in history has society been better prepared for the movement of ideas to bring about profound social transformation," he wrote.[10] He saw the movement of ideas—rather than the usual diplomatic machinations—as the potential catalyst for a lasting global peace. Peace could not be achieved merely through the regulation of territories or the mediation of the inherently tendentious relationships between state governments. Life was becoming more global, thanks to the increasingly intertwined nature of the institutions governing human life and work. This increasing interdependency demanded a new level of coordination beyond the traditional construct of national states.

Otlet was hardly the first European to dream of a united transnational government. In the early fourteenth century, Dante wrote an unpublished treatise in which he called for the formation of a united Europe under a single emperor. Erasmus, William Penn, Jean-Jacques Rousseau, Jeremy Bentham, and Immanuel Kant, among others, each in their own time, described their visions for a unified European system.[11] Most called for a permanent European Congress, an international judicial system, and regulation of commerce—a model not far removed from the present-day European Union. Not until the early twentieth century, however, did a truly internationalist approach to governance seem feasible.

In 1910, Theodore Roosevelt had called for a "league of peace" to prevent war "by force if necessary." President Taft later called for a "League to Enforce the Peace" with Wilson and Henry Cabot Lodge (who would later oppose the League of Nations' covenant). During World War I, writers like Norman Angell and Leonard Woolf gave the idea further momentum in England, where the socialist Fabian Society put forth a proposal based on Woolf's ideas (with an introduction by Bernard Shaw) that would ultimately help shape the draft covenant put forward by the British government.[12]

Up until then, few writers had dared to imagine the possibility of a truly universal—or even pan-European—government. The papacy had once asserted itself as a unifying force on the continent, but that authority had long since faded. The nineteenth century had seen the formation of a tangled knot of political alliances following the post-Napoleonic Treaty of Vienna (1815), which laid the foundation for the political struggles that would usher in the war. These alliances, which had once, paradoxically, been envisioned as a way to preserve the peace, had instead trapped these nations in a web of military obligations that were now plunging the continent into bloodshed.

As war swept the continent, Otlet emerged as a prominent polemicist for the cause of an international league, penning numerous essays and opinion pieces in newspapers in Switzerland, France, and elsewhere. La Fontaine too wrote his own polemics, notably *The Great Solution: Magnissima Carta,*[13] in which he made his case for a new global organization of nation-states. The horrors of the war had kindled in both men a larger vision for the possibilities of international cooperation—one informed by their firsthand experience with building the Union of International Associations, and by their painstakingly methodical approach to classifying and synthesizing the intellectual output of disparate institutional entities.

Otlet recognized that emerging communication networks were shrinking the effective distances between organizations of all stripes—including nations—and that this technological transformation might serve as the catalyst to global geopolitical change. He believed that a new international system might one day emerge, offering the world's peoples "a new life" and a "new course of ideas." The old transnational alliances were becoming outdated, while an increasingly networked system of transport, postal services, and electric signals were facilitating an unprecedented exchange of people, goods, and ideas—all contributing to the formation of a new worldwide economy.

"Through publications, travel, congresses, exhibitions, science, literature and the arts, national and ethnic thoughts are gradually coalescing into universal thought," he wrote.[14] In such an environment, it was all but inevitable that old national and regional political structures would gradually give way to a more unified governing apparatus. To that end, Otlet proposed a world charter designed to embody the principles of human rights and international coordination. It would be founded on the creation of legal and political systems designed to promote peace, cooperation, and good will, and on the preservation and "expansion" of life. Lest this sound typically grandiose and visionary, he also proposed the adoption of the metric system, Greenwich Mean Time, a universal calendaring standard, international signage system, an international language such as Esperanto, and measures to safeguard the international freedom of the press (and an official international journal published by the central federation). Ultimately a worldwide constitution, binding all humanity into a "universal society," would be superimposed atop existing national structures to safeguard the security and well-being of people everywhere. "Life is becoming international," he wrote, and governance should follow suit.[15]

Otlet then proceeded to outline the contours of an organization that would safeguard such an undertaking. He proposed a set of universal human rights—freedom of individual safety, freedom of conscience, freedom of religion, universal suffrage, and the protection of private property—that would serve as the organization's foundational principles. To safeguard those rights, he drafted a detailed charter, enumerating how it would work, including an executive branch, parliament, international court, and a standing army drawn from member nations, with a mutual agreement to defend each other's borders. There would be laws governing international economic relations and transportation, as well as a requirement that fully one quarter of all national budgets, at every level of administration, go toward education and cultural initiatives designed to promote "intellectual and moral development." Similarly, one quarter of the international federation's budget would go toward creating mechanisms to promote the exchange of ideas in science, literature, the arts, and education.[16]

Thus, Otlet saw the project of a world government as closely intertwined with his vision of a global information-sharing network. One would feed the other. A federation of international associations—much like the one he had already established—would play a central role in the enlightened bureaucracy that he imagined. Under the auspices of a world government, his Universal Book and the attendant organization to support it might finally come to fruition. The path to world peace would run right through Brussels.

The End of War also includes language to safeguard the self-determination of nations and races, ensuring their mutually assured protection from conquest. Yet while he mentions the rights of indigenous peoples, he also calls for more advanced states to improve "the moral and material conditions" of such groups within their zones of influence—a veiled reference to Belgian's ostensibly humanitarian mission in the Congo.

By this time the Belgian government had taken formal ownership of the Congo, since Leopold II had transferred control in 1908. Now renamed the Belgian Congo (formerly the Congo Free State), the country had seen slight improvements in its humanitarian situation following the public outcry over the atrocities during Leopold's reign. A new constitution explicitly banned forced labor, and the territories were now governed in part by "traditional chiefs" (*chefs coutumiers*) who presided over the local courts. But the government still banned any explicit political activity and relied on its infamous Force Publique to maintain order (the Force Publique had achieved one of Belgium's few military victories during World War I, capturing part of Germany's colonial territory during the East Africa Campaign). The African colonies had figured as one of the underlying causes of the war; securing their future would be a necessary precondition to a lasting peace.

As the League continued to take shape, the fate of the colonies remained a controversial topic. While there was a vocal, idealistic faction hoping to include language about human rights and the equality of nations, the colonial powers resisted, hoping to protect their economic interests and ensure that their territorial interests would stay profitable. For all of his own high-minded rhetoric about international peace and unity, Otlet struggled to reconcile these noble sentiments with the fact of widespread European colonialism and Belgium's own continued vested interest in the Congo. In *The End of War,* he devoted a section to "international territories" or colonies, asserting that "the African continent is a completely international domain." Echoing Leopold's humanistic rationalizations for his exploitation of the indigenous population, Otlet characterized the entire European project in Africa as "a great civilizing mission," one born of friendship toward the native populations and predicated on a respect for their religious beliefs, family structures, and property rights—and, in the process, "making them appreciate the benefits of civilization."[17]

Africa aside, for the rest of the world, Otlet prescribed a new order, one based on respect for individual rights, protection of nationalities, races, and religions, and a federation of nations to safeguard those rights. Along with La Fontaine, he came to believe that the only antidote to war would be, effectively, a league of nations ("société des nations"). Otlet's faith in the essential goodness of international associations seems naive to us now, given the checkered history of every institution of international bureaucrats ostensibly charged with pursuing the public good, but for Otlet the righteousness of international associations was an article of faith, as symbolized by the logo he proposed for the new entity: an orange sun against a white backdrop, with the Latin inscription: "Per Orbem Terrarum Humanitas Unita" ("Humanity United throughout the World").

Soon after the war started, La Fontaine left for England and later traveled to America, realizing that it would be all but impossible for him to carry out his pacifist work under the German occupation. Otlet decamped for Holland, then Switzerland, where in 1915 he began to lay the groundwork for a Swiss-Belgian association to promote neutrality. He also wrote a series of newspaper articles and made numerous public appearances to promote his vision for an international league. His essay began to attract attention around the continent, not all of it positive. When Otlet tried to leave Switzerland for Paris on April 11, 1916, he and his wife were detained at the border and had their passports confiscated. The French consulate in Geneva had branded him a threat to national security on account of his "dangerous pacifism." The Otlets returned to Lausanne.[18]

The French may have been particularly suspicious of anyone carrying a pacifist message, especially in the midst of the Battle of Verdun, during which 500,000 sons of France would ultimately lose their lives (alongside another 500,000 Germans), triggering a wave of desertions. Otlet appealed the decision, reassuring the French of his intentions. "I have not come to Paris to make propaganda for peace," he wrote in a letter to M. Durand, the Prefect of Police, arguing that "I am an internationalist . . . I am not a pacifist." He went on to explain that whereas the pacifist seeks peace no matter the cost, the internationalist accepts the inevitability of conflict. Pacifism, he argued, often lurches toward short-term peacemaking at the cost of effecting meaningful long-term change. The internationalist, by contrast, is willing to wait for a "peace that will last." To that end, he looks for a new level of international coordination with the requisite checks and balances to address the root sociological causes. "Antagonisms multiply along with the points of contact," he wrote, "and the spheres of friction grow larger." He ended his entreaty on a personal note, invoking his two sons then fighting the war. "If this war does not end in the creation of a stable League of Nations, all of our sacrifices will have been in vain. And, if the fault is ours, our dead will renounce us."[19]

Three months later, the French government acquiesced and lifted the ban. But Otlet chose to continue his campaign from Switzerland, trying to answer his critics and clarify his intentions before finally traveling to France in October 1916. There, he continued to receive a hostile reception, where many traditional socialists strongly opposed the concept of a society of nations. "Who expresses this propaganda?" asked an editorial in a July 1917 edition of the French periodical *La Victoire*. "Foreigners!" the editorial went on, pointing to the work of an unnamed "Belgian pacifist" working in Switzerland.[20]

Otlet continued to press his case elsewhere. In a letter to King Albert, Otlet advocated for strict adherence to the Belgian Constitution, the maintenance of neutrality, and the repudiation of all violent annexation, arguing that Belgium should be "the first of the small nations." As the war drew to a close, he hosted a series of conferences on topics, including the idea of a new worldwide organization of nations.[21] He also worked on a piece called *The International Problems of War*. That essay crystallized Otlet's vision of world harmony. Dedicated to King Albert, the essay sounds a clarion call for peace, grounded in a remarkably precise, logical, and quite grounded, analysis of the causes of war.

"We have reviewed actual events," he wrote. "To catalog the facts, to clarify them, to retain from among them what is essential, to link one to another, to follow them towards more general facts and then to others yet more general still, such has been the task we have proposed if not accomplished."[22] World order would only emerge from the careful scrutiny of facts and appearances, drawn from as many points of view as possible—nothing less than a complete taxonomy of the causes and effects underlying international conflict.

Otlet tried to exert whatever influence he still wielded over the Belgian government to promote his vision, but he met with a lukewarm response from the minister of foreign affairs. "I do not think it is the job of the Belgian Government to take the initiative now to organize a League of Nations," the minister wrote. "The primary issue at the moment is beating our enemies, and all our efforts should be concentrated on this goal."[23] Pressing on, in June 1917, Otlet published a *Global Organizational Plan for the League of Nations,* a slim book that recapitulates many of the ideas from *The End of War,* though with greater force. Now, along with the idealism came mechanisms for enforcement. This organization would also hold legislative, executive, and judicial powers to carry out its will. One of the founding principles of the League held that no state could declare war on another state, and that war would constitute a crime. The states would also have a common defense against outside aggressors and would agree to settle differences between each other through recourse to the judiciary.

The question of Africa came up again as well, with Otlet suggesting that the continent be declared an international domain under the aegis of an African Union. But Otlet's view of the colonial experience had changed. Even he had to admit the Congo experiment had been a disaster and was undermining Belgian credibility. In 1917, Otlet finally came out in favor of ending colonization and promoting independence for the African states. Later that year, he composed an article that appeared on the first page of the *League of Nations,* a journal published by the Boston-based World Peace Foundation. "International war is abolished, just like civil war and private war," he wrote, "and with it are abolished all the consequences it entails: military preparations, hostile alliances, the destruction of lives and goods, the conquest of territories, the domination of populations against their will." He concluded: "All war shall be considered a crime against humanity, and punished accordingly."[24] We can read the experience of the war in the militarism of this antiwar rhetoric.

At forty-eight years of age, the effort behind his activism was beginning to show. Otlet began to endure a crisis of confidence in 1917. His son Jean was still missing. The institutions he had founded lingered in war-torn limbo. And his beloved country remained under German occupation. Worried for her husband's state of mind, Cato enrolled him in psychotherapy. In his notebook, he wrote despairingly of his mental anguish during the war. "My life?" he wrote. "Work, travels, thought, writing, organizations, simple [tasks]." At last, he wrote, "My adventures during the war are difficult to tell about. My wanderings, my difficulties, the state of my ideas, the direction of my action." But he ends on a hopeful note: "My life and my work continue."[25]

Finally, on February 2, 1918, Otlet received confirmation of his worst fear: his son Jean was dead. Filled with grief, Otlet wrote: "Two hours ago, I received the news from Marcel. Poor Jean died a hero... in him, I see the entire crop of young men... and I sense within me, stronger hatred, not against men, against such irresponsible beings, the unknown hand that kills." Otlet resolved to make Jean's sacrifice worth something. He resolved to dedicate the rest of his life to a massive undertaking: "a great human city, completely devoted to Peace."[26]

When the war finally ended, Otlet left Switzerland and returned to Brussels, broken-hearted yet determined to press on. While the war had raged, the IIB had managed to stay afloat, thanks to the efforts

Otlet (center), La Fontaine (far right), and others at the Palais Mondial, 1930s. Reproduced with permission from the Mundaneum.

of a skeleton crew led by its secretary, Louis Masure. Otlet and La Fontaine, who had returned from London, came back to find that their vast bibliographical collection remained largely intact in a city ravaged by war. They secured a meeting with the new prime minister, Léon Delacroix, who lent his support to the restoration of its budget and eventually agreed to reinstate their government subsidy and grant them a large parcel of land in the Parc Woluwe for them to build their long-envisioned international museum, which they now dubbed the Palais Mondial (World Palace).

Housed in a wing of the grand Palais du Cinquantenaire, the Palais Mondial was by April 1921 welcoming 2,000 visitors per day. Upon entering, visitors were greeted by a large sphere, representing world peace, and the Francis Bacon–inspired tree of ages, depicting the evolution of life. They could then proceed through the thirty-six-room display, with each room devoted to a particular country. There were also exhibits devoted to particular

Telegraph Room at the Palais Mondial, Brussels, 1920s. Reproduced with permission from the Mundaneum.

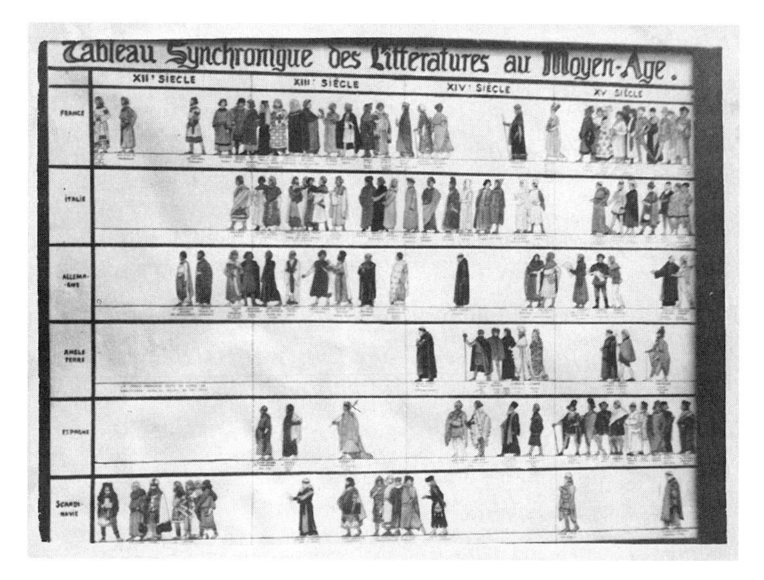

Synchronized Table of Medieval Literature (*Tableau synchronique des litteratures au moyen-age*) at the Palais Mondial, 1920s. Reproduced with permission from the Mundaneum.

Mathematics exhibit at the Palais Mondial, c. 1921. Reproduced with permission from the Mundaneum.

scholarly disciplines—mathematics, chemistry, and paleontology—and rooms devoted entirely to new technologies, such as a sprawling telegraph room and later a large chamber populated with the latest microfilm readers.

Visitors would pass a 1,000-seat lecture hall, surrounded by smaller meeting rooms. Finally, they would come to the heart of the Palais: the library. The library included three separate chambers: one devoted to the Universal Bibliography, with its millions of index cards stored in wooden cabinets; another to the International Encyclopedia, an enormous filing cabinet holding a wide array of articles, pamphlets, and "fugitive literary material"; and third, a collection of books.

Otlet would eventually establish a research service that allowed researchers to submit queries via mail or telegraph. For the modest sum of 27 francs, anyone in the world could send in a request. Otlet's staff would then try to answer the question by poring through the catalog. Within a matter of months, with the world happily pursuing its prewar research and business enterprises, inquiries poured in from all over the world—more than 1,500 a year, on topics as diverse as boomerangs and Bulgarian finance.

Taken together, the operation constituted "a current, ever-expanding repertory of knowledge," as a visiting British librarian named Berwick Sayers described it, "without any of the drawbacks of the encyclopaedia in book form, which is obsolete in many particulars on the day of publication."[27] Despite its popular appeal, however, some visitors seemed less than impressed. An American librarian named William Warner Bishop visited later that year, confessing to his disappointment at the "general impression of inefficiency and confusion."[28] Another American visitor, Ernest Cushing Richardson, gave a more balanced review: "These plans and their authors have been treated by many as grandiose, visionary and unpractical," he wrote, "but the authors of the idea have pegged away for twenty-seven years and have produced for the world of which we are a part, a going concern.... It is true that most of these are not only incomplete but in large part only sketchy. On the other hand at almost every point the material, so far as it goes, is organized in such a way as to be a concrete and permanent contribution.... Even where unorganized in detail there is little that can be called confused."[29] Journalists began to poke fun at Otlet and La Fontaine's grandiose plans. One article in a Brussels paper caricatured them as "the solemn fools of the 'Palais Mondial,'" with a cartoon depicting a small man with thick, heavy glasses waving a piece of chalk in the air. Another periodical skewered Otlet and La Fontaine for trying "to transform the whole of Brussels into a vast city of cards."[30]

While the museum resumed its operations, Otlet continued to focus much of his energy on the cause that had occupied him throughout the war: a league of nations. As the war had drawn to a close, Otlet had also resumed his correspondence with Hendrik Andersen, in hopes of rekindling their plans for a World City that might stand at the center of a new world government. Both men seemed ready to pick up where they had left off, exchanging letters of escalating affection and flattery.

"It seems as though there were a little group of men in the world created with a soul and special insight into human progress," Andersen wrote to Otlet in May 1917, "and, although they may be wandering about apparently in the dark, beneath their mantle they hold one may almost say a sacred light to illumine and guide future progress."[31] Similarly lofty prose would characterize Andersen's letters to Otlet over the years to come. And while Otlet rarely indulged in such self-aggrandizing rhetoric, Andersen's flattery seems to have struck a chord.

Otlet, somewhat as James had a few years earlier, came to see the World City as something of a castle in the sky, on the order of Plato's Republic, Saint Augustine's City on a Hill, Francis Bacon's New Atlantis, and of course Thomas More's Utopia. Yet, unlike James, he also saw the World City as more than just a literary construction. Like Andersen, he intended to get it built. In addition to describing its philosophical underpinnings, he tried to outline the financial and political systems that would be required to bring the World City to life, drawing extensively on the planning work he had put into his proposals for the League of Nations.[32] Indeed, he came to see the World City as part and parcel of the League of Nations, the physical manifestation of the world government of which they had both dreamed.

On September 29, 1918, Andersen wrote excitedly to Otlet, sharing his optimism about the league's prospects. "God grant that a League of Nations after the war may inspire faith and confidence in those who desire to establish a World City to administer international affairs justly," he wrote. "President Wilson stands for the divine rights of freedom for all nations and expresses his views so strongly toward democracy and international justice that it is for us to take courage and lay the sacred foundation stones for building a City to God and Humanity."[33]

During this same period in the latter days of the war, President Wilson and his chief collaborator, Colonel House, had conceived a secret think tank of 150 scholars, geographers, statisticians, mapmakers, economists, lawyers, and journalists—including the famous *New Republic* editor Walter Lippman—to ponder the future of the postwar world. Cryptically named "The Inquiry," the group met furtively in a conference room at the New York Public Library on Fifth Avenue. President Wilson had taken a lead role in promoting the cause of the League in the United States. Although the idea of the League had originated entirely in Europe, Wilson's stature and devotion to the cause brought him forward as its primary publicist. When victory came a few weeks later, America seemed poised to lead the world into a hopeful new postwar era, with a strong League of Nations designed to ensure a lasting peace.

"Hurrah for the victory," Andersen wrote to Otlet, one week after the final armistice on November 11, 1918. "Now is the time to move our project, and we must do it in a definite and political way."[34] Andersen had already written a letter to Wilson's adviser Colonel House. Now he suggested that Otlet try to make contact directly with House in Paris in advance of the upcoming Paris Peace Conference, to persuade him of the merits of including their proposal for a World City as the centerpiece of the new League of Nations, where twenty-nine heads of state would gather to consider adopting the covenant of the new organization.

"This is the time to work on the project," Otlet wrote Andersen. "What are you doing? What do you suggest? You should come here, now."

A few days later, Andersen replied. "I beg Senator La Fontaine and yourself to attempt the impossible and present the City project at the Peace Conference, so that it can be built at Tervuren. Keep this project alive!"[35]

At the Paris Peace Conference of January 19, 1919, the attendees endorsed the covenant of the League of Nations. Otlet and La Fontaine both attended, as did their friend Paul Hymans, who served as Belgium's principal representative. At their behest, Hyman introduced a resolution calling for the establishment of an organ to coordinate international intellectual relations, but the League failed to adopt it. Otlet and La Fontaine tried to introduce another, similar resolution a few weeks later, but without success.[36] Otlet's hopes for the League of Nations were quickly meeting with disappointment. Although his "Declaration of the Rights of Nations" had served as a foundational document, as did "War and International Conflict," the organization that ultimately emerged out of Paris bore little resemblance to the one he had originally envisioned. When the nascent League chose Geneva over Brussels as its headquarters in April of that year, he took the news poorly, calling it "that lamentable decision." There would be no World City in Tervuren.

Instead, the League would make its home in Switzerland. "Few countries are less cosmopolitan," Otlet complained in a letter to his friend, the Liberal Party politician (and later second president of the League of Nations) Paul Hymans. With the memory of his recent exile there presumably fresh in his mind, he dismissed the entire country as "a big hotel," a place "where nobody stays."[37] Otlet felt the decision to choose Switzerland was also symptomatic of the League's deeper shortcomings: namely, a tendency toward pragmatic, bureaucratic deal making that failed to serve what he saw as the organization's higher purpose. "Neither the problem of war, nor of armaments, nor the manufacture of ammunition nor the problem of economic relations, have been squarely addressed," he wrote. "Those who presume to lay the foundations for a new order to ensure world peace, have succumbed unconsciously to the occult forces that perpetuate conflict."[38]

Still, Otlet continued to press his case. On June 14, 1919, he sent a letter to President Wilson, upon whom he bestowed the title of "Patron" of the Union of International Associations.[39] Otlet and La Fontaine had also convinced King Albert to grant the Union government recognition, a status that extended to the Palais Mondial, making it a quasi-governmental institution. Thus they hoped to exert some influence over the direction of the League of Nations by fostering a formal relationship between the League and the Union.

The Palais Mondial—which Otlet and La Fontaine ran as a small duchy in the Palais du Cinquantenaire—received renewed government assistance, while La Fontaine, one of Belgium's official representatives to the League who also happened to serve as both vice president of the Belgian Senate and secretary general of the Union of International Associations, continued to look for ways to promote the Union's interests and align it as closely as possible with the League's evolving organizational structure.

Otlet and La Fontaine's hopes for the League received a further setback in November 1919, when the U.S. Senate refused to ratify the covenant. Despite President Wilson's enthusiastic support, the Republican-controlled Senate, heavily influenced by Wilson's political nemesis Henry Cabot Lodge, had refused to ratify membership on the grounds that it would compromise U.S. sovereignty. To encourage this sentiment, Lodge inserted language to weaken the bill before Congress, arguing against the League's required commitments to mutual defense and an international court. Wilson barnstormed the country relentlessly to generate popular support for the League. After traveling 8,000 miles in twenty-two days, he succumbed to a debilitating stroke that forced him to cut the tour short. A few days later, the Republicans prevailed as the Senate failed to ratify the Treaty of Versailles.

The U.S. decision not to join the League all but assured the organization's ultimate failure. As the fledgling League took shape, the remaining members squabbled over policies and procedures, with the Great Powers—particularly Britain and France—asserting their dominance and securing permanent spots in the League's governing council, while the lesser powers were forced to accept a reduced role in the General Assembly. Otlet's friend Paul Hymans emerged as the primary spokesman for the smaller countries, making an impassioned plea for them to play a sustained larger role. After the inevitable rounds of diplomacy and politicking, the League began to look less and less like a fledgling world government than a tempestuous teacup of bickering bureaucrats. However, there was some room for optimism. When Otlet received news that the League had included an article about organizing institutions, he wondered whether he might yet play a role.

Soon after the British diplomat and Scottish peer Sir Eric Drummond took office as the League's first secretary-general, an office he would hold from 1919 to 1933, Otlet and La Fontaine began lobbying him to have the League recognize the Union of International Associations, giving it international legal status and funding. Eventually, they hoped to expand the scope of the UIA and the Palais Mondial into an international entity on a par with the League itself. Drummond gave them a polite hearing, replying diplomatically, "The International Secretariat will be only too glad to work with an organization in the furtherance of the cause of peace and internationalism."[40] In September 1920, Drummond paid a visit to the newly constructed Palais, where he spent at least an hour touring the exhibits, and by all accounts seemed favorably disposed to the project. In March of the following year, however, Otlet and La Fontaine found their proposal to the League rebuffed, not even by Drummond but by one of the secretaries, who wrote, "Your scheme of intellectual organization strikes me still as a little in advance of time." The League had more pressing priorities to attend to, but might perhaps be ready to revisit their proposal in three or four years' time.

Otlet, though now growing accustomed to disappointment, responded with the impotent rage of the rejected. He had clearly expected a very different outcome, and he recognized the League's decision as the possible death knell for his fondest dream. Throwing aside his usual diplomatic caution, Otlet wrote back to the underling who wrote the letter, unleashing a volley of antibureaucratic rage. "Your letter will mark a date in our history. It is a disappointment, a great disappointment . . . the great rock that we have tried to get to the top of the mountain has been pushed back down. . . . Alas!" He went on to rail against the failure of the assembly to rise to the historic occasion. "Faced with all these great intellectual challenges the Council has become indecisive. How saddening to witness the spectacle. . . . How typical this is the Supreme Council, which will continue to govern human affairs by force, by ruse, and in secret, and all for the benefit of the privileged."[41]

The mixed reactions to the Palais Mondial, coupled with the League's rejection, began to take their toll on Otlet's psyche. "Everywhere I meet the same situation," he wrote plaintively in a letter to his friend George Lecointe of the Royal Observatory. "I am discarded, or eliminated . . . my fellow countrymen do not understand me."[42] Now entering his middle years, having lost his son to war, he felt as if he were losing steam. His progressive enthusiasm, born of a more hopeful period before the war, seemed increasingly out of touch with the cautious and pragmatic bureaucratic atmosphere that seemed to envelop the continent. Otlet's Left-leaning politics also left him out of favor with the increasingly conservative Belgian government, and his single-minded determination—which had proven so critical to his early success—now proved a liability as he failed to adapt to the changing political conditions. "He had begun to feel that changes in Belgium were passing him by," writes Rayward, "that he had slipped from that position of eminence in the social and intellectual life of Brussels which he had occupied before the war."[43]

Making matters worse, Otlet had attracted the ire of many Belgians when he offered to host a Pan-African Congress at the Palais Mondial from August 31 to September 2, 1921. The congress gathered

Pan-African Congress at the Palais Mondial in Brussels, August 31–September 2, 1921. Reproduced with permission from the Mundaneum.

black activists from around the world, welcoming delegates from the fledgling National Association for the Advancement of Colored People (NAACP), England's African Progress Union, and representatives from other like-minded organizations around the world. The first Pan-African Congress had been instigated by NAACP founder W. E. B. Du Bois and taken place in 1919 in Paris, where initially it met with opposition from the French government.

The second congress provoked even more controversy than the first. Du Bois advocated that Belgium relinquish its entire colony in the Congo, along with 9 million native residents, to a newly created African international zone. Du Bois had first proposed his scheme before the first Pan-African Congress, proposing the creation of a new "Africa for Africans," created out of the former German colonies in Ethiopia. "Within a generation, young Africa should know the essential outlines of modern culture," wrote Du Bois. "We can, if we will, inaugurate on the dark continent a last great crusade for humanity." Du Bois also appealed to European self-interest. "With Africa redeemed, Asia would be safe and Europe indeed triumphant."[44] The settlement that Du Bois envisioned would welcome black residents from across Europe and the Americas who wished to resettle. It would encompass 20 million inhabitants, spread out across 2.5 million square miles of the African continent, and be administered by a governing body that would unite local entities while providing support for education, cultural initiatives, and social reform. "It is quite Utopian," wrote one contemporary observer, "and it has less than a Chinaman's chance of getting anywhere in the Peace Conference, but it is nevertheless interesting."[45]

When the Congress organizers approached them, Otlet and La Fontaine extended an eager welcome, offering them the Palais Mondial as a venue. Otlet undoubtedly found Du Bois's vision appealing, insofar as it accorded closely with his own preferred solution to the African problem: namely, an end to the colonial project, coupled with the repatriation of displaced Africans back to their ancestral homeland. And Du Bois's notion of "social progress" for Africans must have resonated with Otlet's own positivist views of cultural evolution.

Before the Congress was convened, Otlet had offered several suggestions to the organizers. The primary one was that "negroes alone" should organize the conference and determine which white attendees they should include. He also suggested an agenda, voicing his hope that the Congress would address a series of political and theoretical questions about the black experience, including history, ethnography, sociology, culture, and economics, and questions of rights and governance. Finally, he suggested that the Congress "carry out a scientific examination of all the great questions which concern black people and to establish to this end an international center devoted to study and documentation, one which should report to the Congress on the one hand and on the other hand to other great organizations."[46] He closed with an offer to house such a center in the Palais Mondial.

In 1920s Brussels, Otlet's support for the Pan-African Congress marked him as a progressive, if not even a radical. But he also shared in the widespread racial prejudice of the time. Fourteen years later, he would reveal his views on race in a brief passage in his book *Monde*, in which he characterized the Negro brain as "less developed than in the white species." And while he advocated for the "unification of races" in principle, he nonetheless pointed to what he saw as their distinguishing racial characteristics. "For intellectual work, they have little aptitude, but they excel in dancing, fencing, swimming, horse riding, and all physical activities."[47] Such views, though they may strike us as repellent today, were certainly widespread among white Europeans of the era. And Otlet's public demonstration of support for the cause of African self-determination put him well out of step with most of his countrymen.

Otlet and La Fontaine met with vehement public opposition over the Pan-African Congress. Perhaps this should come as no surprise, given Belgium's economic interest and financial investment in the Congo. At a gathering in London in 1921, many of the attendees had signed a statement condemning Belgium's colonial experiment in the Congo and declaring their commitment to future reform. At the first Congress in Paris, the delegates had further affirmed their belief in the "absolute equality of races" and in "denying the God-appointed existence of super-races, or of races, naturally and inevitably and eternally inferior." Such sentiments raised hackles among the Belgians. Some news outlets began to conflate the efforts of the organizers with the more radical agenda of the Marcus Garvey–led Universal Negro Improvement Association—which actively sought to push Europe and the United States out of Africa—while others saw the Congress as a communist plot to stir up the Congolese natives against their occupiers. A correspondent for the Antwerp newspaper *Neptune* noted that it was reported in the United States that the organizers had received "remuneration from Moscow (Bolsheviki)." "The association has already organized its propaganda in the lower Congo, and we must not be astonished if some day it causes grave difficulties in the Negro village of Kinshasa, composed of all the ne'er-do-wells of the various tribes of the Colony, aside from some hundreds of labourers."[48]

The flurry of public outrage spurred attendance at the event, so much so that white Europeans outnumbered the black participants by a wide margin. But there were nonetheless some notable black activists in attendance, including Du Bois, who wrote a report on the Congress for the NAACP's signature publication, *The Crisis*. His article noted the high turnout and the crowd's seemingly keen interest in the topic of African liberation—which came as something of a surprise, given the relative scarcity of black people in Belgium at the time. "It was not long before we realized that their interest was deeper, more immediately significant, than that of the white people we had found elsewhere," he wrote. "Many of Belgium's economic and material interests centre in Africa in the Belgian Congo. Any interference with the natives might result in an interference with the sources from which so many Belgian capitalists drew their prosperity."[49]

The Congress featured a series of speakers, mostly African Americans as well as a few educated native Africans and a handful of white European men and women. Their presentations discussed the state of affairs for black residents in a wide range of countries, including the United States, France, England, Libya, and, naturally, the Belgian Congo. Despite the generally hopeful and laudatory rhetoric about brotherhood and the betterment of the race, the meeting was also marked by dissent. A schism arose between two broad groups, with the American and British black attendees forming one bloc, opposed by their French and Belgian counterparts. The issue at hand came down to passing a series of resolutions coming out of the conference. On the one hand, the American and British attendees wanted to pass a strongly worded resolution condemning the exploitation of colonial Africans. But the French presiding official balked, instead stewarding through passage of a less vigorous resolution calling for the establishment of research institutes to study the state of black affairs in each colonial power. The author of that resolution was Paul Otlet, who hoped such institutions would ultimately participate in the vast worldwide network of institutions that he had long envisioned.

The barrage of negative attention to the Pan-African Congress wore on Otlet. Public resentment bubbled over. In the meantime, Otlet and La Fontaine's influence suffered a setback with the resignation of Prime Minister Delacroix, who had supported their efforts in the crucial years immediately following the war. The new government under Prime Minister Georges Theunis proved much less receptive to their utopian ideas. In February 1922, Otlet received word of the government's decision to use the space occupied by the Palais Mondial as a commercial exhibit hall; it now seemed less and less likely that the plans for a new Palais Mondial in the Parc Woluwe would ever come to fruition. Otlet scrambled to rally international support, convening a gathering of supportive scholars and diplomats for another Quinzaine Internationale. Representatives from sixteen countries (including the United States) took part, but none of the major European powers sent a representative. Nonetheless, Otlet presented a resolution to the conference calling for recognition of the Palais Mondial as an international organization—rather than a

Belgian one—funded by contributions from individual governments. The resolution also called for formal recognition from the League of Nations. The attendees dutifully endorsed the proposal, but it proved a toothless contract, as no further action was ever taken. The Belgian government later refused to acknowledge the agreement, arguing that it had not been formally represented (despite the presence of La Fontaine, by then vice president of the Senate). The League of Nations took delivery of the resolution but also neglected to act. When the League finally called for a long-awaited meeting on intellectual exchange between its member nations—the very cause that Otlet had lobbied the institution to embrace—neither he nor La Fontaine was invited.[50]

In July 1923, Otlet took a further blow when he received news that he would have to clear everything out of the Palais Mondial to make room for a rubber fair. The site of international cooperation he had once championed would now give way to the rubber industry—the basis of King Leopold II's colonial misadventures in the Congo. While the government would perhaps have forced the UIA to move regardless, the irony was nonetheless brutal. Otlet tried to appeal the government's decision, while exploring other alternatives to preserve his collection and the institutions that had taken shape around them. He considered moving the entire operation to New York, Paris, Rome, or the Hague. He turned to his friend Patrick Geddes for support and even briefly considered joining forces with Melvil Dewey's son Godfrey to create a new home for the collection. But in the end these efforts came to nothing, and the government moved forward with its plans. At one point it offered Otlet the opportunity to move the contents of the Palais to a series of abandoned railroad sheds. He refused. La Fontaine used what remaining influence he had to lobby on the institution's behalf, eventually helping to locate more suitable lodgings in a small building on a less-traveled boulevard that could at least be used for storage purposes.

On the morning of February 12, 1924, a team of men gathered to begin removing the contents of the Palais. Otlet refused to cooperate, ordering the presiding government official to leave immediately and barricading the doors with file cabinets. The government men proceeded to storm the building, breaking in and tearing apart the collections. A visiting reporter compared Otlet to Marius weeping before the ruins of Carthage.[51]

8

Mundaneum

The turmoil surrounding the Palais Mondial affair left Otlet deeply embittered but no less determined to see his projects through. Although the Belgian government eventually relented and agreed to let him return some of his material to its old quarters, they withheld any financial or administrative support, leaving Otlet and a tiny staff to begin the slow and painstaking work of reconstructing the collection. Frustrated by his dealings with the new Belgian government and increasingly despondent over the prospects of returning the Palais Mondial to its former glory, Otlet began to focus his energies elsewhere.

But all was not lost. Just as the Belgian government had withdrawn its support, a glimmer of renewed interest came from the League of Nations. The Swiss national librarian Marcel Godet, working on behalf of a League-sponsored committee, wrote a report evaluating the possibilities for collaboration between the League and the IIB. The report leveled some harsh criticisms—namely, "that it lacked clarity, a critical sense, and that it endeavored to embrace every country, every language, every period, and every subject, a task which it is difficult to achieve. The Institute undertook one gigantic task after another." The report went on to criticize the IIB for its seemingly haphazard approach to collection development, its reliance on the Decimal Classification, and, finally, "its propensity to overrate the value of index cards."[1] But Godet nonetheless praised the IIB's hard work and long track record and suggested a path toward partnership with the League. Godet proposed that the League commission the IIB to develop a comprehensive catalog of authors drawn from all the world's libraries, as well as a more in-depth catalog of works on particular topics like bibliography and intellectual cooperation.

Recognizing the offer as a lifeline and determined to make an impression on the League, Otlet convened a joint conference of the IIB and the UIA in September 1924, scheduling it to take place in a location conspicuously close to the League's new headquarters in Geneva. In preparation for the conference, Otlet wrote a long manifesto detailing his views on possible international collaborations among the IIB, the UIA, and the League.[2] His proposal went well beyond the realm of bibliography. Unsurprisingly, the document is imbued with positivist rhetoric, calling for a "restoration of spiritual power" that would emerge with the grand synthesis of knowledge of which he had so long dreamed. For Otlet, that synthesis would manifest on three levels: as an idea, an institution, and—importantly—as a physical establishment. He dubbed this new entity the Mundaneum.[3]

"It is necessary," he wrote, "that by intellectual vigor we achieve a Science, an encyclopedic synthesis, a science of the universal, embracing everything we know." This would, in other words, be far more than a list of authors' names. The project he proposed would ultimately take shape as "an organization for relations between men, and for their relations with things, an organization which should be oriented towards synergetic action, which takes into account at one and the same time and as a whole, all men, all countries, all relations: the Earth, Life, Humanity."[4] Building on his earlier work with the UIA and the World City proposal, he called for a new quasi-governmental entity that would coordinate the activities of a far-flung network of international and national associations, groups and individuals. Here, more carefully and forcefully articulated than ever before was Paul Otlet's dream of a global intellectual bureaucracy, one that was both physical and spiritual, providing administrative services as well as an architectural infrastructure for displaying the combined intellectual output of these organizations. This was surely more than Godet had bargained for when he suggested that the IIB might produce a simple bibliography.

The conference attracted participants from about seventy organizations, including the League. Melvil Dewey's son Godfrey chaired the proceedings, which also included Dorcas Fellows, the new editor of the Dewey Decimal Classification. Together, they came to agreement with Otlet and La Fontaine on one important outcome: the long-sought reunification of the American scheme with the Universal Decimal Classification. Further resolutions were passed, most of them formalizing the recommendations in Otlet's manifesto. In the wake of the conference, the UIA presented the League with a proposal for an expanded relationship, including its participation in the League's Committee for Intellectual Cooperation and the adoption of an international statute legitimizing the UIA's work.[5] If successful, this proposal would mark an important step toward integrating the organizations. Otlet and La Fontaine soon began making plans to move their associations' headquarters to Geneva.

Despite the warming of relationships, the League proved slow to move forward. Just as Otlet was preparing his proposal for the UIA and IIB, the French government had been working on its own proposal to create a new home for the League's Committee on Intellectual Cooperation in Paris. Unlike Otlet, the French could offer the financial support of their government. Months of bureaucratic maneuvering ensued, as Otlet and La Fontaine tried to curry favor for their own proposal or to find an accommodation with the French. They also tried to shore up their support across the Atlantic, leaning on their connections with Godfrey Dewey to drum up support from deep-pocketed American philanthropist organizations like the Carnegie Corporation and Rockefeller Foundation. Months turned to years, as members of the various committees, institutions, and governments involved in the process struggled to come to terms. The reality of international cooperation was proving far more difficult to achieve in practice than the idealized picture that Otlet had painted in his proposal.

Now nearing sixty and undoubtedly worn out by his disappointments with the Belgian government, the glacial deliberations of the League, and the bureaucratic infighting that continually seemed to stymie his projects, Otlet began to see his reputation deteriorate. The Swiss librarian Godet—who had resurrected Otlet's hopes for cooperation with the League—now called for a more "competent man" to direct the negotiations. In a letter to a member of the International Institute on Intellectual Cooperation, he shared his misgivings about the IIB's leaders: "I know that their utopian tendencies do little to inspire confidence in those who hold the purse strings."[6] By 1926, the relationship between the League and the IIB seemed once again to be foundering.

Never one to shrink from adversity, Otlet pressed forward as best he could. In 1926, the IIB issued an ambitious new edition of the UDC. By this time Otlet had ceded much of the work on the classification scheme to an energetic young Dutchman named Donker Duyvis, who in 1921 had taken on a role as secretary of a committee to update the classification and seemed to inject the effort with fresh enthusiasm. He also began work on planning yet another international conference, to be held in 1927 at the

Mundaneum (as the reconstituted Palais Mondial had now been renamed).[7]

During this period, Otlet returned once again to his ideas about the World City. He had kept in regular contact with Hendrik Andersen over the years, trying to keep the embers of their shared ambition alive. But Andersen was starting to grow disillusioned and seemed less and less interested in pursuing the support of the League. "I have no faith in the organized 'League of Nations' as it stands today," he wrote to Otlet on March 20, 1928, "and I do not want to except [*sic*] any proposal or any assistance from this diplomatic and political international body of retired politicians." "No! I will have nothing to do with them."[8] By this time, Otlet had already begun to cast about for new partners to help move the World City project forward.

In 1927, Otlet made contact with Le Corbusier, the French-Swiss artist and architect whose building designs would help define twentieth-century modernism. Born Charles-Edouard Jeanneret, he had changed his name to Le Corbusier in 1920—adapting an old family surname, but also making a personal choice that reflected his conviction in the liberating possibilities of severing one's ties with the past.

Le Corbusier made his reputation by promoting a new, industrialized architectural style rooted in the scientific management theories of Taylor and Ford. Like Andersen, Le Corbusier had long harbored an interest in designing whole cities, having laid out an ambitious design for a 3-million-person metropolis in 1922—the Ville Contemporaine, or "Contemporary City." At one point he even went so far as to propose demolishing most of central Paris and replacing it with a vast grid of sixty-story towers. Sympathetic to the industrialism and scientific management principles of Taylor and

Ford, Le Corbusier envisioned an entirely new kind of city, one better adapted to the exigencies of modern life. Cities had a choice to make, he famously said—"architecture or revolution."[9] He held out hope for the possibilities of large-scale central urban planning, which he thought might provide the key to preserving peace amid the social unrest that had accompanied the spread of industrialism throughout the European World.

Le Corbusier, as it happened, had recently pitched his own proposal to the League of Nations, for a new headquarters building in Geneva—a competition that attracted no fewer than 377 architects. His design called for a complex of flat structures housing more than 500 offices, as well as a great assembly hall raised on stilts, nestled between the lake and the mountains and surrounded by an English garden. In the strikingly modernist style that was becoming Le Corbusier's signature, the assembly hall would be covered in polished granite, with large windows set between thin steel mullions. It was to be a sleek, forward-looking monument to the new world, devoid of the labored symbolism and flourished excess of the old.

Le Corbusier's proposal won over the jury, who chose it as the winning entry. But it failed to pass bureaucratic muster when the French juror—whose countrymen's submissions had not been selected—complained that Le Corbusier's plan had not been submitted using the required China ink. The jury agreed to reopen the competition, triggering a round of behind-the-scenes maneuvering that Le Corbusier later characterized as "the blackest state diplomacy." In the end, the jury awarded the project to a rejiggered proposal by two Frenchmen, an Italian, and a Hungarian. A French academic architect named Henri Paul Nénot was also named director general of the project. Le Corbusier accused the winning team of plagiarism—and later tried to sue the League for copyright infringement—but it was too late. He had lost the commission.

Disappointed by the decision but encouraged by a strong reservoir of support from several of the state ambassadors who had favored his original proposal, Le Corbusier was all too ready to accept Paul Otlet's invitation to work on a new proposal to the League. Otlet and Le Corbusier shared an idealistic streak and an appetite for ambitious, world-changing projects. The two men seemed to hit it off from the start. Indeed, Le Corbusier soon began referring to Otlet as "Saint Paul."[10]

In 1928, the two men published a pamphlet called *Mundaneum*, consisting of a general description of the project written by Paul Otlet and six drawings accompanied by a detailed narrative from Le Corbusier.[11] The Mundaneum that Otlet had been describing in idealistic terms now took on a more precise architectural form. Here at last was the synthesis and apotheosis of everything that he had been struggling for over the course of four decades, centered around the bibliographical project with which it had all begun.

Appropriately, the pamphlet proposed a World Museum as the centerpiece of the project, envisioned as an enormous prism containing metal walkways, shelving, slides, and elevators throughout the space, all contained in a vast glass enclosure. The entire structure would "stand straight and smooth, with its silent walls surmounted by a crown glittering windows of its upper part." The structure would also house an International Library, including the Universal Bibliography, which Otlet suggested could be decoupled from its institutional parentage in Brussels and made a permanent resource of the League of Nations. Finally, the structure would also include a university with numerous lecture halls, carrying forward the educational mission that Otlet had conceived for the old Palais Mondial.[12]

In a way reminiscent of Geddes's Outlook Tower, visitors would climb to the top of the structure—where they could enjoy panoramic views of the Alps—before descending the structure through a long winding passage, similar to Frank Lloyd Wright's famous spiraling pathway through Guggenheim (built in 1959), only in reverse. They would move through successive exhibits representing the stages in humanity's development (echoing the positivist view of progressive cultural development). In the first stage they would encounter material about prehistory and humanity's shared roots. Moving even further down the spiral, they would encounter all the world's great civilizations. As they wound along the spiral, they would have the opportunity to explore a wide range of modern civilizations in further detail. "The map of the world grows and changes, pulsating like a flower photographed in slow motion."[13]

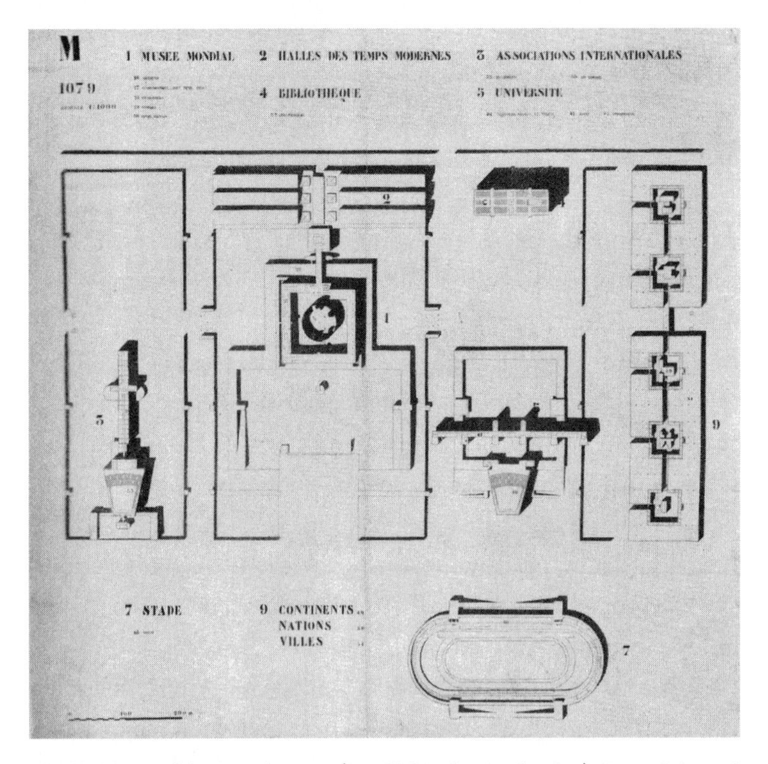

The Buildings of the Mundaneum (*Les édifices du Mundaneum*). From Otlet and Le Corbusier, *Mundaneum* (1928).

The entire complex would consist of five major components:

- **The International Associations Building**: A sprawling office complex to host the headquarters of various international associations, along with a meeting hall to accommodate up to 3,000 spectators.
- **The Library**: A universal library with an enormous card catalog modeled on the Universal Bibliographical Repertory, along with additional lecture halls.
- **The University**: An international university intended to accommodate at least 500 students that would build close relationships with other universities all over the world.
- **Exhibition space**: Publicly accessible spaces with room for permanent and temporary exhibits on a wide range of topics.
- **The World Museum**: A museum divided into three parts: Works, Time, and Place.

While Le Corbusier is best known for his highly functional designs, he also harbored a strong interest in the mystical. His designs for the Mundaneum include a number of occult symbols, including a large so-called Sacrarium, a smooth cylindrical enclosure intended to gather and channel spiritual energies, within which stood stone figures of the "great initiates in which humanity, through the ages, had imbued its mystic power."[14] The Sacrarium may well have been influenced by the teachings of the Theosophical Society, the quasi-Buddhist school formed by the Ukraine-born spiritualist Madame Helena Blavatsky. The term "great initiates" likely stemmed from a book of the same name (*Grands Initiés*) by the theosophist writer Edouard Schuré.[15] Early in his career, Le Corbusier also wrote of his pilgrimage to a house designed by the noted theosophist architect J. L. M. Lauweriks, where he found the ornate, spiritually inspired geometrical figures deeply affecting, influencing his lifelong propensity for modular design.[16] Otlet too had developed an interest in theosophy. The conference room in the Palais Mondial featured a monumental painting of Prometheus by the theosophist John Delville, who also served as secretary of the Brussels Theosophical Society. In 1923, Otlet gave a lecture to the group (most likely at Delville's invitation), discussing the spiritual aspects of the World City project; three years later, he addressed a larger gathering of 2,000 theosophists in Ommen, the Netherlands, where he encountered Krishnamurti, the group's anointed spiritual leader.[17]

Just as the theosophists hoped to reveal the eternal knowledge of the ages, the Sacrarium would serve as an architectural vessel to reveal humanity's mystic power and "the hidden idea of the spirit of history itself."[18] Otlet threw himself into the spirit of it all and described the Sacrarium as a place that would locate the common ground among all human philosophies, "an expression of the supreme relation of the self to the Universe of the Totality, the Infinite."[19] Separated from the worldly knowledge of the rest of the Mundaneum, the Sacrarium would provide "a constructed ritual in which one first breaks away from the profane space in order to discover the sacred."[20] This would lead to a great "Union of Spirits," bringing the realm of the spirits into contact with the terrestrial world. The Sacrarium represents perhaps the pinnacle of Otlet's ambition: from trying to collect a comprehensive bibliography of books and articles to unifying all of the world's religions is a leap of nearly mind-blowing proportions.

For Otlet, the years of disappointment and frustration had led not to a scaling back of ambition, but to its very opposite. Le Corbusier undoubtedly played a catalytic role. Of all the people with whom Otlet had made a connection, and with varying degrees of success—Dewey, Geddes, Andersen, and so forth—Le Corbusier was perhaps the only one with the practical training to act upon them. He not only drew up plans; he had an established track record of getting his projects built. Still, the fact that he and Otlet were willing to propose such an ambitious scheme speaks to the spirit of optimism, universalism, and international kinship that still lingered in the air toward the end of the 1920s, and in the wake of the formation of the League of Nations—a last gasp of internationalism before the continent would once again succumb to the nationalist fervors that led to World War II. The pamphlet tried to fuse Otlet's long-standing dreams of the pooling of knowledge with the political aims of the League of Nations. "From anywhere on the planet, the image and the total meaning of the world can be perceived and understood," he wrote. He hoped that the Mundaneum in this incarnation would become "a sacred place, inspiration and fount of great ideas and noble activities . . . one element of the great epic and wonderful adventure continued through the ages by mankind."[21]

While the Mundaneum embodied the utopian internationalism that enjoyed so much currency in Europe in the early twentieth century, it was hardly the first proposal for such a "harmonious synthesis" of setting and learning. Throughout history, librarians and architects have understood the delicate relationship between knowledge and physical architecture. The old Library at Alexandria—the center of learning in the ancient world—was also built with Aristotle's ideal of peripatetic scholarship in mind, with wide colonnades and open spaces to encourage scholars to walk with each other and engage in dialogue about whatever mutual topics of interest might arise. Similarly, Melvil Dewey saw his scheme as not just a classification system, but a physical map of human knowledge, specifying how the physical layout of libraries should reflect the conceptual order of the decimal classification.

Otlet often spoke of his bibliographical work in architectural terms as well. Otlet and Le Corbusier shared remarkably similar language in their works on architecture and the organization of knowledge, using terms like "plan," "standardization," and "synthesis" to characterize their respective visions of an idealized future.[22] Otlet saw parallels between the modernist revolution in architecture and the possibilities for reinventing the book, which he at one point described as an "architecture of ideas." In the years that followed, architecture would play an increasingly important role in Otlet's thinking about organizing the world's information. Indeed, he came to see the Mundaneum as both an architectural plan and a metaphor for a new, enlightened form of civilization.

Building on the Mundaneum proposal, Otlet and Le Corbusier began work on a formal proposal to the League for a new version of the World City. Their proposal took its inspiration from Andersen's original plan but now put Otlet's Mundaneum at the heart of things, placing a much stronger emphasis on creating a physical space to support the great positivist project of unifying all human knowledge: "a symbol of global intellectual unity for all of humanity."

Echoing Andersen's earlier work, Otlet recalled the ideals of the ancient world, of Egyptian temples, the Oracle at Delphi, and the Library of Alexandria, places where "great souls realized great works of thought and writing." He then went on to invoke the medieval heritage of cathedrals and universities of Bolognae, Paris, and Oxford, "where intellectual work took place amid admirable architecture," and the Renaissance, where grand palaces and great cosmopolitan capitals provided space for the encyclopedic ambitions of the British Museum (founded in 1753) and the Institut de France (1795), as well as the great world's fairs that had provided an unprecedented degree of public access to the latest cultural, intellectual, and scientific advances. "Everywhere," he wrote, "are great works of the mind and monuments to the human spirit."[23]

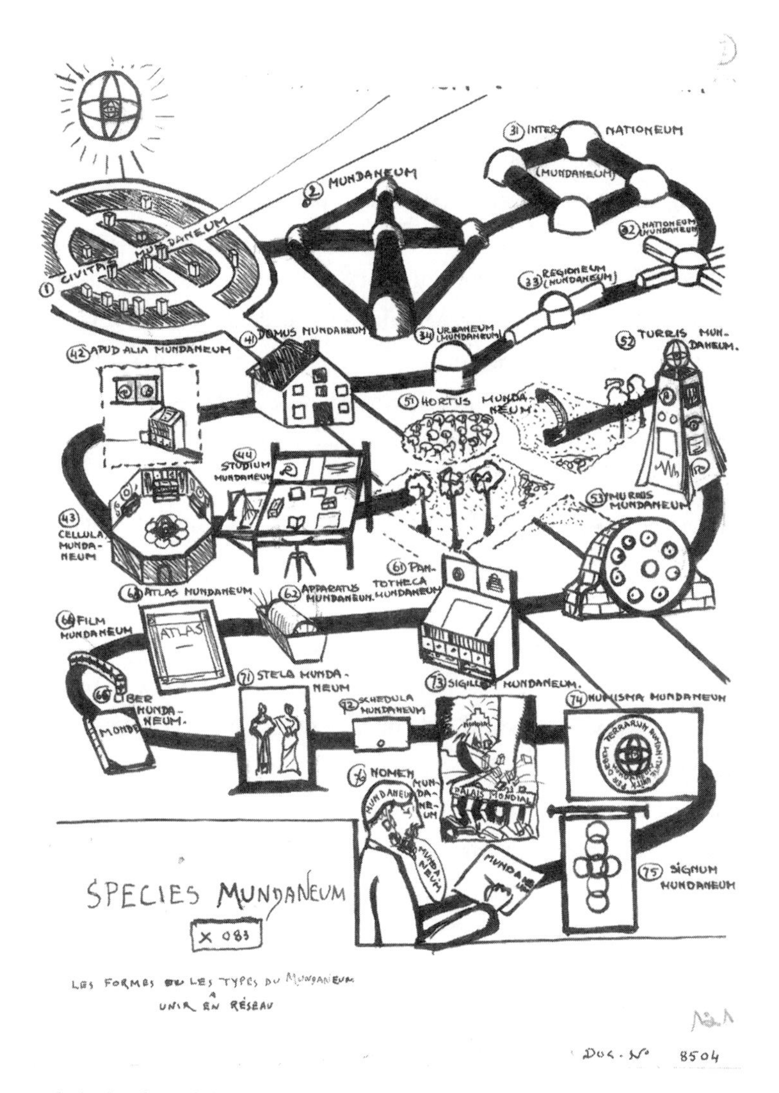

Otlet's plan for a global network of institutions linked together with the Mundaneum, 1930s. Reproduced with permission from the Mundaneum.

In 1929 Otlet published a new pamphlet entitled *World City*—obviously borrowing heavily from Andersen—addressing it directly to the officials of the League of Nations. The proposal was startling in its grandiosity, even for Otlet. Here he proposed a site encompassing nearly 1,400 acres, costing an estimated $50 million (in 1929 dollars). The city would act as the nexus and central transmission point for a vast intellectual network that would radiate outward across the world, with the Mundaneum at its center, serving not simply as a repository of words and images, but as an active force for propagating knowledge and fostering shared cultural understanding. A federation of museums would also play a critical role in extending the reach of the Mundaneum network into the everyday lives of citizens across the world, by transforming the contents of the collection into usable material that interpolated words and images in new ways.

When Andersen got wind of Otlet's partnership with Le Corbusier, he was incensed. In a letter dated June 4, 1929, he wrote, "I am surprised, that you should drift on to take other proposals and support other projects, when you have had in your own hands for so many years my project for a World Center of Communication, upon which I have given my life, energy and fortune," he wrote. "I must tell you that I am surprised and cannot in any way sympathize with the corruption of my original and organic plans."[24] His indignation is perhaps understandable. After all, Otlet had taken the germ of Andersen's vision and effectively handed it over to another architect. "Why should you, dear Friend, pick my plans and life's work to pieces?" However justified Andersen's pique under the circumstances, he seems to have undermined his credibility when he also accused the late President Wilson of having stolen his idea for the League of Nations in the first place—a baseless claim that pointed only to Andersen's growing megalomania. In 1926, he had presented his plans to Mussolini; a few years later, he would take part in a radio broadcast hailing Mussolini's achievements as harbingers of the World City to come.

Otlet parted company with his former collaborator. Ignoring Andersen's opposition, he threw himself into the new plans, presenting the project to the president of the League's Assembly and staging a public exhibition at the Villa Bartholoni in Geneva beginning on August 1, 1929. Otlet made appeals to private donors and foundations. He placed high hopes that the United States—perhaps Rockefeller—would help to underwrite his project. Finally, Otlet thought, there might be an opportunity to fund the project through widespread donations collected via a subscription offering, or what we might today call crowdsourcing. The proposal received favorable reactions in the Swiss press, which welcomed him as a visionary. "Men like him are needed," went one editorial. "They are the whip that shakes our laziness, our inertia, our tendency to accept things as they are and ignore what should be."[25]

At last it looked as though the World City project might once again be gaining traction. But Otlet's timing, as ever, was not propitious. The Wall Street crash came just a few weeks later.

Recognizing that the spiraling worldwide depression had dashed any realistic hopes of persuading the world's governments to finance a massive new construction project for the League of Nations, Otlet shifted his focus away from the World City and began to explore other projects. Over the years he had kept up his correspondence with Patrick Geddes, as the two men evolved their thinking about museums and other cultural institutions. Now, he began to consider the educational potential of emerging new technologies like television and radio. "Visualization on the screen will become a fundamental teaching method," he predicted.[26] To accomplish that pedagogical goal, he placed a great emphasis on what today we might call instructional media: charts, tables, diagrams, and other kinds of graphic depictions that anticipated the multimedia displays we now all take for granted. Otlet had long held an abiding interest in the graphic presentation of information. He envisioned a wide range of audiovisual teaching tools—interactive textbooks, bibliographies, and syllabi—facilitated by emerging technologies like gramophones and slide projectors, as well as microfilm, radio, and cinema. Taken together, these devices would begin to function as "auxiliaries to the teacher, the extension of the word and the book."[27]

These ideas marked an important departure from the traditional approach to museums that prevailed across Europe and North America in the late nineteenth and early twentieth centuries. Most

Exposition of Documentary Material, Paris (1946). Reproduced with permission from the Mundaneum.

museums still served primarily as collectors of objects, trying to assemble unique or exemplary works of art, craft, or specimens of the natural world. Museum curators tended to resist the temptations of populism, refusing to cater to what they saw as a working-class lust for entertainment.[28] Few museum curators saw themselves as active producers or distributors of knowledge; they were in the business of preservation. But by the early twentieth century, a few visionary museum exhibit designers like Joseph Urban and Stewart Culin had started to rail against the dry and oppressive character of most museums. Envisioning a more democratic, accessible approach to curation, they found inspiration in other public spaces like department stores, theater sets, and hotels.[29] By the 1920s, the twin forces of modernism and popular culture were starting to challenge the traditional role of museums as highbrow arbiters of culture.

Otlet's thinking about museums was evolving too. While his early efforts with the Palais Mondial reflected an indiscriminate zeal for collecting everything he possibly could, that approach had run aground with the loss of government support for his collection. He began to think about alternative ways to make the museum collection available to a wider, geographically distributed public—not the physical artifacts themselves, but the intellectual capital they represented. In 1925, Otlet described what he called the Encyclopedia Universalis Mundaneum (EUM). Initially this was to be an encyclopedia stored on microfilm that would allow subscribers to purchase topic-specific collections of content drawn from the archives of the Palais Mondial. A collection of thirty-eight microfilm images about Egyptian civilization would cost fifteen francs; forty-eight images about Jewish civilization would cost nineteen francs; while twenty images about the "barbarians" would cost a mere eight francs. At the high end of the scale, a grand tour of the entire world, consisting of eighty-seven images, would cost thirty-five francs.[30] Thus, anyone in the world could access the contents of the museum, from anywhere in the world.

He also envisioned a companion project called the Atlas Universalis Mundaneum (AUM), a collection of diagrams on large movable boards, intended to present a visual synthesis of material about a large number of topics, grouped into history, geography, and science. The AUM would draw on and synthesize the contents of the Mundaneum for use in a classroom, museum, or other exhibition space. Otlet intended for the material to be updated constantly by experts.

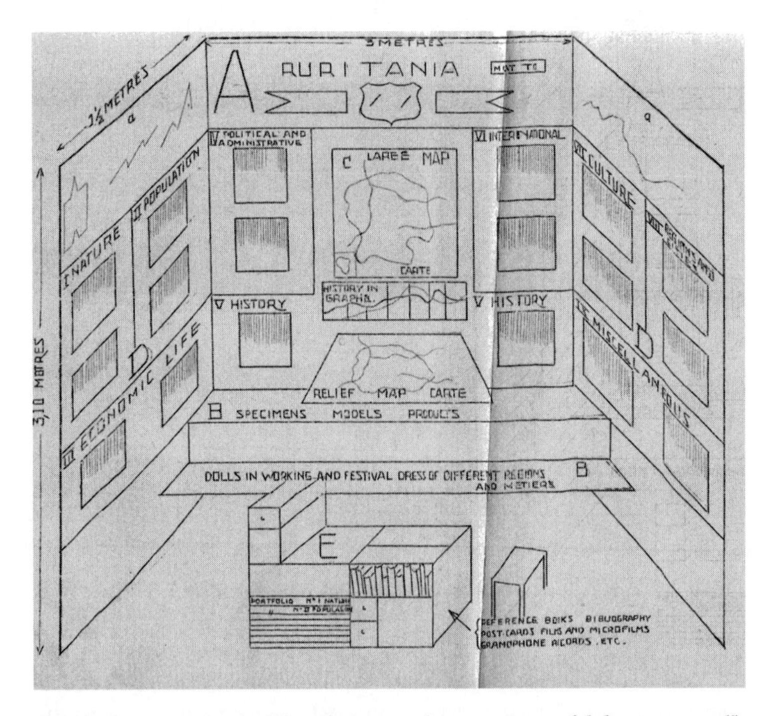

Sketch to accompany the "classification and presentation of didactic material"; from Otlet's unpublished *Encyclopedia Universalis Mundaneum* papers. Reproduced with permission from the Mundaneum.

Instead of presenting a single, authoritative curatorial view of the collection, the encyclopedia would allow users to customize their own interactions with it. "Everyone will become his own editor," Otlet wrote in one of his unpublished papers, by using a system consisting of three basic functions: (1) a "print layer" allowing for anyone to contribute a piece of work to the collection; (2) microfilm, allowing its contents to be stored efficiently, then reused in the form of lectures, screen display, or reproduced on paper; and (3) the Mundaneum, which would provide a networked repository of documents that could then be indexed, cataloged, and widely distributed—and from which any individual user could put together a collection of material customized around a particular set of interests.[31]

In short, he envisioned the EUM as an interactive knowledge space that would incorporate aspects of a traditional encyclopedia—entries listed according to alphabetical organization and cross-referenced—along with a set of multimedia displays, all tied together through an underlying classification system (the UDC). The encyclopedia would consist of indexed written entries, while the atlas would offer images that could be printed out on posters and hung from the walls, each displaying some combination of charts, illustrations, diagrams, and textual content to provide an overview of a given topic. This fusion of word and image would offer a new kind of encyclopedic experience, taking shape in what Van Acker calls a "museological space."[32]

Information was no longer fixed and could not be presented in a static form. The new encyclopedia would do more than merely serve as a traditional storehouse of written articles; it would have a dynamic function, acting as an active learning center and catalyst for social change. Otlet and Geddes believed above all in improvement, progress, and education, holding that the museum could play a role even more important than traditional books or classroom teachers.

But Otlet saw education as self-education, and therefore as extending beyond classrooms. To that end, he made heavy use of sometimes cartoonish illustrations designed to appeal to a broad spectrum of ages and levels of educational accomplishment.[33]

Inspired by Geddes's visualization work, Otlet began to collaborate with an artist named Alfred Carlier on a series of graphic posters, providing overviews of particular domains of knowledge that were connected to the International Museum's collections. Otlet's collaboration with Carlier dated back to 1920, when they worked together on the handmade historical dioramas for the Palais Mondial. Otlet called his collaborator the "learned artist." Over the years that followed, Carlier would produce thousands of such plates, which would form the basis for the encyclopedia and for a subsidiary collection that came to be known as the Atlas of Civilization. Otlet also marketed the material as instructional aids to educational institutions, such as the Montessori school in Amsterdam.[34]

These visual representations look very much like modern information graphics, anticipating the fusion of word and image that we now expect to find on websites, apps, and other screen-based multimedia. Otlet also proposed the notion of a standard "muséothèque," a kind of syndicated exhibition kit consisting of prepackaged display material, appropriate for use in museums or schools. Such exhibits could then be reproduced, combined in new ways, and distributed to other museums or cultural institutions all over the world.

Otlet's concept of syndicated museum displays owed a large debt to Otto Neurath, the Viennese philosopher and sociologist. The two had met while preparing for an exhibition in Geneva in 1928, in which Otlet presented his Atlas of Civilization. Neurath had come to Geneva to exhibit the visualizations he had created in his capacity as director of the Vienna-based Museum of Society and

Economy. He had founded the museum in 1925, with a mission to educate the working classes about issues of economics, health, education, and any number of other topics relevant to their daily lives: from unemployment and alcoholism to tuberculosis and the value of sports. The museum marked an important departure from typical museums of the day, insofar as Neurath saw the museum in a growing competition for attention with the attractions of film and other forms of popular entertainment.[35] Whereas traditional century curators resisted such popularization, Neurath embraced it as a positive social good. "Modern man receives a large part of his knowledge through pictorial impressions," wrote Neurath, observing the proliferation of photographs, lantern slides, daily newspapers, and film that continued to proliferate throughout modern society.[36] He believed this emerging visual culture offered enormous potential for improving the quality of human communication. He proposed a new approach to "pictorial statistics" that eliminated the discursiveness of written language and opened up the possibilities of communication to a much larger audience of people from anywhere in the world. As he put it in his oft-quoted slogan: "Words separate, pictures unite."

Neurath saw the growing public appetite for visual stimuli as good news for museums. "Exhibitions, museums are thoroughly products of this drive to watch," he wrote.[37] Rather than focus on the display of individual historical or cultural artifacts, he wanted to create a new kind of museum devoted to the display of useful information, to help Viennese workers understand the larger forces that were shaping their lives—echoing Geddes's socially minded intentions with the Index Museum. Neurath not only "collected facts," as Neurath scholar Nader Vossoughian puts it, but he constantly looked for new ways to portray them. At one point he developed a system of "statistical hieroglyphs"—called

ISOTYPE—intended to provide a universal visual language for conveying numerical information. The idea was to create a visual system that could transcend differences of language and culture, while creating a clear and compelling tool for juxtaposing related concepts. Adapting the Futura font that had been made popular by the Bauhaus movement—which embraced the clean and geometrically proportioned sans serif font for its modernist aesthetic—Neurath set about designing images that evoked the same kind of elegant simplicity. The images were intended to be modular, so that two signs could be combined to form a new one. For example, the sign for "shoe" (an image of a shoe, logically enough) could be added to the sign for "works" (a building with a smokestack) to denote a shoe factory.[38] These pictographic language associations shared an important trait with Otlet's Universal Decimal Classification: the ability to juxtapose one concept with another, by means of a semantic link.

After first encountering each other's work at the Geneva conference, Otlet and Neurath recognized their shared interest in creating graphical representations of large bodies of knowledge. Neurath's sophisticated visual language for museum spaces appeared to offer a perfect complement to Otlet's encyclopedic ambitions. And while Otlet spoke little German and Neurath little French, they nonetheless discovered an intellectual and spiritual kinship. "[Your] determined chase after all-encompassing international plans made a deep impression on me," Neurath wrote. The two men agreed to cooperate formally in 1929, creating a joint venture between Otlet's Palais Mondial (still a functional entity for administrative purposes, though closed to the public) and Neurath's Vienna-based Museum of Society and Economy. The entity they formed was to be called the Novus Orbis Pictus (invoking a 1658 pictorial encyclopedia published by the Czech bishop and educational reformer Johann Amos

Comenius). Neurath's museum would take responsibility for designing the exhibits, while Otlet's World Place would collect and classify the material. The Museum of Society and Economy would develop "pictorial statistics and the systematic cartographic designs," as Neurath put it, while Otlet's organization would work toward "comprehensiveness and systematization in the early stages of work."[39] These pooled intellectual assets would form the basis for exhibits to be created according to a standard set of specifications. "Thus the World Museum will consist primarily of combinations of reproducible pieces," read the agreement, which went on to elaborate the group's plans for "consolidating scientific research, as well as spreading and democratizing knowledge and its instruments to disseminate social education throughout the world."[40]

Otlet, still with the Index Museum in mind, had at first envisioned the Palais Mondial as a self-contained physical entity. Neurath opened up a new line of thinking. He envisioned replicating the contents of his own museum, then making them available to a much broader public. Such a syndicated approach, Neurath believed, could mesh well with Otlet's aspirations. Inspired by Otlet's original plans for the Palais Mondial, Neurath hatched an elaborate scheme to create museum exhibits that could be easily replicated and installed in museums anywhere in the world. The museum experience was a moveable feast. He believed in expanding the scope of museums beyond their traditional role as "curiosity cabinets" of unusual physical artifacts, usually limited to a particular geography or culture. Instead, by reproducing and syndicating their exhibits, museums could widen their audience and democratize the culture at large. At one point Otlet mused about the possibility of taking exhibits out on tour by automobile; Neurath replied that he too had imagined just such a project, equipping a car with a portable projector and cinema display, a "Wandermuseum with car."[41]

Neurath also felt that museums should take advantage of new technologies to apply the lessons of the assembly line to the creation of cultural artifacts. "To speak of the museum of the future is like speaking of the automobile of the future," he wrote in 1933. "The idea that every museum ought to contain unique exhibits comes to us from the past." Just as books were now produced in factories, Neurath argued, so the techniques of mass production (or reproduction) should allow for the creation of portable museum displays, going on to reference Otlet's proposal to create standard copies of museums for display. Such an arrangement would not only yield obvious economic benefits, but would also make it possible to promote a shared sense of history and cultural understanding across international borders. "In the future," he wrote, "museums will be manufactured, exactly as books are today."[42]

In order for mass-produced museums to take shape, however, museum administrators would have to cede a degree of power and authority to their audiences. Instead of relying on elite curators to assemble a collection of objects they deemed important based on their individual expert judgment, the museum of the future would rely on the active participation of visitors to shape the exhibits. "Why must the poor visitor to a museum of natural history look at hundreds of birds, even though he can perhaps hardly distinguish the differences, just because some particular ornithologist considers it necessary?" Instead, he wonders whether a more user-centered museum might put the information on display into a context more useful to the viewer. "Would it not be more important to tell the people whether there are many or few animals of this or that kind, which of them are edible, what the skin of this or the bones of that beast might be used for?" Here Neurath anticipates one of the great cultural debates of today's networked age: the tension between top-down institutional authority and bottom-up user involvement.

For now, the institutions remained firmly in charge, yet Neurath wondered whether the winds of technological change might ultimately force a transformation. "Isn't it curious," he wondered, "we are constantly told that we are living in the age of technology, and yet when we enter a modern museum of natural history, there is no sign of it."[43]

Ultimately, Neurath and Otlet envisioned an environment in which the museum became fully abstracted from its physical expression. Just as librarians had started to draw a distinction between the "book" (the physical volume) and the "work" (the intellectual contents expressed therein), he imagined that museums could take ethereal form—what today we might call multimedia—that could then be replicated in multiple physical locations.[44] The museum was to become, in other words, something like a piece of software.

After their initial meeting in Geneva in 1928, the two men agreed to collaborate in designing a network of museums—a "link of Mundaneums," as Otlet put it—that would draw on a common set of assets, such as a global encyclopedia; an atlas of geographical, social, and economic maps; and specialized publications. Neurath thought one promising avenue might involve persuading individual museums—especially local museums and school museums—to host Mundaneum sections within their collections, which could then be administered from afar in close coordination with the hosting institution.[45] They communicated with museum curators from the Chicago Museum of Science and the German Werkbund as well as the Museum of the City of New York and the Committee on the Association of American Museums, all of which expressed interest in collaborating.

The partnership between Neurath and Otlet proved short-lived. Although they had hoped to solicit funding from the Carnegie Foundation, the 1929 stock market crash put an end to those hopes.

Both men also came to recognize fundamental differences in their approaches to the encyclopedia, with Otlet focused more on the problems of managing a large, encyclopedic body of information, and Neurath on the development of a universal visual language to express a manageable set of facts.[46] Although both men continued to share a conviction in the social value of visual media forms that could transcend national boundaries, their work ultimately took them in different directions. Neurath distanced himself from the Mundaneum, focusing instead on building a publishing business (now dubbed Orbis), leaving Otlet to forge ahead on his own.

Otlet eventually began to work on an even more ambitious work that would consist of more than 8,000 illustrations, attempting to provide a visual orientation to a vast corpus of knowledge drawn from numerous disciplines. This project—to be called the Atlas Encyclopedia Synthetica—was designed with Neurath in mind but never came to fruition. Nonetheless, over the next few years Otlet did begin to produce collections on microfilm known as the Encyclopedia Microphotica Mundaneum that would eventually encompass hundreds of images and charts drawn from the Mundaneum collections.

Otlet, Le Corbusier, Geddes, and Neurath, like the Renaissance encyclopedists, hoped to foster social progress by re-imagining the mechanisms by which people came to understand the world. They also shared a belief in the possibility of conveying information through visual and spatial mechanisms, and taking advantage of the latest industrial technologies to speed the flow of knowledge all over the world, all in hopes of ushering in a new age of interconnected consciousness.

For Otlet, in the end, the effort centered around his beloved catalog, along with its associated encyclopedias and institutions. There would be not one Mundaneum but many, and each one would serve

Exhibit at the Palais Mondial, 1920s. Reproduced with permission from the Mundaneum.

as "a world in miniature, a cosmoscope allowing one to see and understand Mankind, Society and the Universe."[47]

Today, such language may invite ridicule, but Otlet's ideas about museums and encyclopedias revealed an unshakeable faith in the transformative power of intellectual collaboration that evolved in tandem with his thinking about the future of books, encyclopedias, and other emerging forms of media. He saw them all as interlocking parts of a unified whole, moving in all dimensions and beyond.

As Otlet continued to expand on his ideas in writing, his real-world projects continued to suffer. By this point, his work on the Universal Bibliography had started to decelerate. Only half a million new cards were added from 1927 to 1931—representing a far slower clip than in its heyday. Work continued to move slowly on rebuilding the Palais Mondial. In April 1930, he wrote an article in the Palais Mondial's eponymous journal, apologizing for the poor condition of the facility. It was "in a bad state of repair and the upkeep leaves a great deal to be desired," he wrote. "The Management offers its apologies."[48] Even so, he kept trying to rally support—encouraging visitors to petition the government, and even trying to launch a Young Friends of the World Palace organization, a kind of intellectual Boy Scouts—but these efforts too came to naught.

At the IIB, Otlet and La Fontaine had also been gradually displaced from their duties by a new generation of leaders, including the Dutchman Donker Duyvis. These men did not share Otlet's enthusiasm for a central repository; rather they focused their efforts on updating the classification system to be used by local libraries. As the Universal Decimal Classification began to gather more traction after its translation into English, the institute shifted focus away from the Universal Bibliography—much to Otlet's dismay. In July 1932, he wrote a distressed note to La Fontaine, bemoaning the "tendency to put the spotlight on the Decimal Classification, one of the elements of the Institute, and leave in shadow or silence the other elements, notably the *Bibliographia Univeralis* and the Universal Bibliographic Repertory on cards."[49] He also bemoaned the lack of movement at the League of Nations, which had failed to take up his proposal. He singled out Duyvis for condemnation, attacking his competence at rewriting parts of the Decimal Classification. The grand scheme he had spent so many years trying to develop was, he felt, coming apart. Otlet vented his frustration to La Fontaine and to Duyvis, whom he increasingly seemed to view as a primary villain. In one four-month period, he sent no fewer than twenty-six separate letters to Duyvis, taking issue with his various decisions.[50] By 1932, after a series of recriminations against Duyvis and others, Otlet had effectively ceased to play an active role in the IIB.

Now sixty-five, Otlet was physically worn out, mentally drained, and financially bereft. He began to accept that he would have to cede power to the rising generation of leaders. In 1933, he altered his will to exclude the institute from his estate. Otlet and La Fontaine held on to their titles as secretaries-general, until eventually they were "kicked upstairs" and given the title of honorary vice presidents.

Otlet and a small band of supporters continued work on the IIB, pressing forward with the goal of continuing to develop the Universal Bibliography. Committees were formed, processes were codified, and drafts were circulated among various regional secretaries. Duyvis assumed the position of secretary; he would play an instrumental role in the IIB for years to come. Nonetheless, the 1930s had brought a cascade of misfortune. The Belgian government had expelled the Palais Mondial; the League of Nations had failed to live up to Otlet's hopes—failing to prevent or even protest either the Italian conquest of Ethiopia or Japan's 1931 attack on Manchuria, and proving themselves largely the toothless collection of bureaucrats that Andersen (among many others) made them out to be. And the stock market crash ended his cherished dream of building the World City.

On June 1, 1934, the Belgian government delivered a final blow: It would shutter the Palais Mondial once and for all, to make space for the Royal Museum of Art and History. Otlet reacted with predictable outrage, spending the entire day staging a silent protest outside the locked doors. He had not lost his flair for the dramatic. Borrowing a magnificent-looking chair from one of his neighbors, he received the government employee sent to inform him of the expulsion order, surrounded by his supporters. He listened to the edict being read out—"By the order of the ministry of public works . . . no one is allowed to enter"—and then resolved in protest not to leave his chair.[51] When Jean Capart, the director of the Royal Museum of Art and History, was interviewed about the incident, he gently mocked Otlet for his obstinacy. "If King Leopold II asked Paul Otlet for a plan to reform the entire Universe, Monsieur Otlet would respond: 'Sire, you will have it in three days!'" Nonetheless, Capart wouldn't be budged from his plans to force out Otlet, whose "private enterprise" had been "hosted in a gracious and precarious state" for long enough, and the space was needed for a "normal expansion of the museum."[52]

Finally accepting the inevitable, Otlet issued an announcement. "Following deplorable circumstances," it read, "the offices of the Headquarters and General Secretariat of the IID have had to be transferred from the World Palace to 44 Rue Fetis, Brussels."[53] The address was Otlet's house.

9

The Collective Brain

The image of the lone genius remains one of the most durable cultural myths in the Western world. We tend to consider the Da Vincis, Brunos, Galileos, and Copernicuses as maverick thinkers possessed of otherworldly inspiration who fought valiantly for their visions in an unappreciative world. Yet history suggests that great ideas often emerge simultaneously in more than one place and from more than one person. In the late nineteenth century, for example, a number of inventors were independently exploring the possibilities of automobiles, motion pictures, and powered flight. Centuries earlier, more than one printer was experimenting with movable type before Fust and Gutenberg perfected the technology. Paul Otlet was scarcely alone in imagining something like a global information network.

After the German chemist Wilhelm Ostwald won the Nobel Prize in 1909, he poured most of his prize money into founding the Bridge (*Die Brücke*), an organization devoted to improving the worldwide enterprise of scholarly work. Along with his partners, Adolf Saager and Karl Wilhelm Bührer, Ostwald imagined a global scholarly environment that would act like a telephone exchange, fostering connections between like-minded researchers. The Bridge would not only put a given scholar in touch with his or her peers, it would also serve as a kind of central nervous system that would "unite his field of work with every other field." Over time, thought

Ostwald, it might evolve into "the great organism of the entire intellectual world."[1]

Ostwald knew of Otlet's work and drew on it extensively, using strikingly similar language to describe the operation, calling it the "brain of humanity" (Otlet would later use the term "mechanical collective brain").[2] While he planned to use many of the same techniques as Otlet and La Fontaine's Universal Bibliography—collecting information from a broad range of published sources, extracting and transcribing the data onto standard-sized cards, and classifying them using the Universal Decimal Classification—he also favored a slightly different vision for the Bridge, envisioning an entity that was more than just a bibliographical clearinghouse but, rather, an active center of knowledge production in its own right.

Drawing on a central repository modeled on the Universal Bibliography, Ostwald and his partners hoped to produce collections of cards that would provide a comprehensive overview of available knowledge on any specific topic. To accomplish that, the Bridge would need to do more than build a large-scale catalog; it would also have to assemble the scholars to work on it. To that end, he envisioned creating a universal directory of scholars, "containing the addresses of all living knowledge workers," along with pointers to the particular kinds of knowledge those individuals had produced. It would act as a contemporary social network in which both people and documents would serve as nodal points.[3]

Like Otlet, Ostwald also proposed establishing a transnational organization to coordinate all this. His approach differed from Otlet's in several important respects, however. For one, he believed that scholarly communication should take place by means of a global language—a notion that harkens back to John Wilkins's Universal Language and would later find expression in Zamenhof's 1887 proposal for Esperanto. Ostwald and his partners even issued a manifesto in Esperanto (*La organizado de la intelekta laboro per la Ponto*).[4]

Ostwald also believed in using the methods of advertising to propagate scholarly work. Up until then advertising had served primarily commercial interests. Ostwald argued that it could be pressed into the service of scholarship and education, helping to provide a platform for popularizing scientific findings and connecting the general public and the scholarly community. "The engineer cannot talk," he said, advocating that schools should put a special emphasis on ensuring better communication and what today we might call presentation skills. Moreover, knowledge institutions had to make their resources more freely available. "It is not enough to found libraries," he wrote in a 1911 article for *Scientific American*. Creating tools to help nonspecialists navigate the vast spectrum of scholarly thought would be key. To achieve that vision, it would be necessary "to instruct those eager for knowledge in the best methods of utilizing their treasures. And this is by no means as easy as it sounds!"[5] Ostwald's efforts toward realizing a global brain hinged on four key principles: order and standardization; the monographic principle (breaking the contents of books down into their component parts); the popularization of scientific knowledge; and the automation of intellectual work by way of bibliographical institutions and "machines."[6]

The Bridge published a number of pamphlets and for a brief period seemed poised to work. As with so many of Otlet's schemes, however, Ostwald's proved short-lived. Donations from wealthy patrons failed to materialize, and as Germany prepared for war in 1914 the Bridge closed its doors after a three-year run. An effort to resurrect the project after World War I came to naught. The Bridge did leave the world with one lasting legacy: the international standard for paper sizes (like A4) that are still widely recognized across the metric system–using world.[7]

While the Bridge may have faltered, Ostwald's scientific work continued unabated. He generated an astonishing scholarly output of 45 books, more than 500 articles, and 5,000 reviews over the course of his long career—contributing more than his share to the problem of information overload that he had once hoped to solve. Along the way he mentored countless young scholars.

One of those students, a young Russian-Jewish doctoral student named Emanuel Goldberg, went on to take Ostwald's vision in a new direction. After completing his Ph.D. under Ostwald's supervision, Goldberg played an important role at the Zeiss Ikon camera company, where he helped launch one of the first home movie cameras (the Kinamo) and developed a novel "microdot" technique for capturing data at extremely high resolutions—able to store up to fifty complete Bibles in one square inch of film.[8] While, as we saw earlier, photographers had experimented for years with microphotography, by the 1920s the practice was becoming widespread. Banks were microfilming their checks for record-keeping purposes; and the emerging technology was quickly catching on with other businesses and government agencies. Libraries too were recognizing the enormous potential cost savings in collecting microfilm rather than paper. With Goldberg's innovative techniques, Zeiss Ikon recognized an opportunity to explore new ways of managing that information using sophisticated processing machines: what today we might call workstations.[9]

Using the Kinamo moving-picture camera and Goldberg's microdots, the company developed a technology that could place large amounts of data onto microfilm. But locating any given record on a long roll of microfilm still required painstaking manual work. In 1927, Goldberg came up with a solution: a mechanical search engine. The Statistical Machine, as he called it, could search and retrieve a record stored on microfilm by using a mechanical indexing technique. The system allowed for each document stored on film to be marked with descriptive information (or metadata) that could be detected using a combination of photoelectric cells and digital circuits. A user could simply punch in codes on a keypad to create a "search card" (similar to a punch card) that would retrieve documents stored on the film. Goldberg eventually added the ability for Boolean searches (i.e., combining search terms using AND/OR logic), counters to record the number of times a document had been retrieved, and even the use of a telephone dial to enter queries: the world's first dial-up search engine.[10]

Goldberg first patented the device in Germany in 1927, then, a year later, in England and the United States, where IBM's chief scientist, James W. Bryce—who had worked closely with Herman Hollerith on some of the company's earliest punch-card technology[11]—promptly snapped up the rights (Goldberg had transferred the patent to his employer for the grand sum of $2.00). In the mid-1930s, IBM was building its portfolio of electronic devices (even before it had started manufacturing any of them), long before Vannevar Bush, then dean of engineering at the Massachusetts Institute of Technology, published his famous essay "As We May Think." Today, most computer science historians have characterized Bush's Rapid Selector as the first electronic information-retrieval machine. When Bush tried to patent his invention in 1937 and 1940, however, the U.S. Patent Office turned him down, citing Goldberg's work. And while there is no evidence that Goldberg's invention directly influenced Bush's work, Donker Duyvis—Paul Otlet's eventual successor at the IIB—did tell Bush about Goldberg's invention in 1946.[12]

Despite his considerable achievements, Goldberg remains all but unknown today. As with Otlet, his contribution disappeared into the Cold War hall of mirrors that enveloped Europe in the wake of World War II. In 1933, just two years after Goldberg had unveiled his machine, members of Zeiss Ikon's Nazi-controlled Workers' Council stormed into his office and at gunpoint marched him out in the rain to a local bar, where he was forced to stand at attention in front of a Swastika for several hours. They then proceeded to take him into some woods where they tied him to a tree. The next day, they forced him to write his resignation letter from Zeiss Ikon and exonerate his kidnappers. Goldberg soon left with his family for Paris and in 1937 moved permanently to Palestine, where he spent the rest of his days working in relative seclusion. He never resumed work on the Statistical Machine.

Before leaving Paris for Palestine, Goldberg gave a paper at the World Congress of Universal Documentation, a conference organized by the International Institute of Documentation (as the IIB was now called) and held at the Trocadéro from August 16 to 21, 1937. The conference focused largely on the potential application of microfilm for information storage and retrieval, with a series of speakers—including Otlet, who according to one observer made "magnificent improvisations"—exhorting the participants from forty-five countries to take advantage of the emerging technology to facilitate the global exchange of scholarly information.[13]

Also in attendance at the conference was the Prussian State Library director Hugo Krüss—who three years later would supervise the Nazi inspection of the Palais Mondial—leading a large delegation of twenty high-ranking German officials to ensure that Nazi interests were represented in any international agreements that might emerge from the congress. He gave a presentation imposingly titled "The Domination of Knowledge."[14]

The conference also featured a much-discussed speech by H. G. Wells, the English science fiction writer who for the past several years had been developing his own ideas about what he called a "world brain." With war clouds gathering, Wells urged the scientists, librarians, and publishers on hand to focus their attention on how their efforts might best prevent armed conflict from arising. "All the distresses and horrors of the present time are fundamentally intellectual. The world has to pull its mind together, and this [Congress] is the beginning of its efforts. Civilization is a Phoenix. It perishes in flames and even as it dies it is born again. This synthesis of knowledge upon which you are working is the necessary beginning of a new world."[15]

A dedicated socialist and pacifist, Wells believed that humanity would have to move to a stateless future, under the rule of a single world government and a class of self-selected technocrats—or "samurai," as he called them—to guide its progress. Only when the conflict between nation-states had been eliminated could humanity finally realize its spiritual and intellectual potential. Worldwide dissemination of recorded knowledge was an essential step along that path. Like Otlet, Wells believed that better access to information might help prevent future wars.

Beginning with his 1905 work, *A Modern Utopia,* Wells had developed a fascination with the problem of information retrieval—the need for better methods for organizing the world's recorded knowledge. This led him to reject old values and institutional strictures and embrace a mechanistic approach, one founded on Taylorist ideals of scientific management and a belief in the power of science to solve humanity's problems, and the coming war in particular. The political and economic imbalances that were leading to war resembled diseases attacking a body beset by a compromised nervous system. By improving the flow of information, humanity could restore its collective moral, political, and intellectual health.

In 1939, Wells published *World Brain*, a collection of essays and lectures drawn from his decades of thinking about the possibilities of new technologies for strengthening humanity's collective intellect. "Encyclopaedic enterprise has not kept pace with material progress," he wrote, arguing that the advent of modern telecommunications and photoreproduction technologies had made "practicable a much more fully succinct and accessible assembly of fact and ideas than was ever possible before."[16] Wells saw universal access to knowledge as more than just an intellectual boon but as a crucial step toward an uplifted society, one that "foreshadows a real intellectual unification of our race." In one startlingly prescient passage, he wrote: "The whole human memory can be, and probably in a short time will be, made accessible to every individual. . . . It need not be concentrated in any one single place. . . . It can have at once the concentration of a craniate animal and the diffused vitality of an amoeba."[17]

Wells and Otlet were very likely aware of each other's work—both men attended the same 1937 conference, at which Otlet delivered an impassioned plea for universal access to human knowledge—but there is no evidence that the two men ever met. Still, the parallels in their thinking suggest that both men were drawing on the same wellspring of ideas and were certainly kindred intellectual spirits. Both had pronounced internationalist impulses, calling for the establishment of a single unified world government. Both saw the outpouring of published information and the proliferation of institutions of higher learning as positivist manifestations. In time, as Wells put it, a "super-human memory" would engender an expansion of humanity's mental faculties. Rampant inefficiencies in the governance of human affairs precluded people from living up to their collective potential. A more unified, coherent system of access to knowledge, therefore, would ultimately lead to world peace. "Without a World Encyclopedia to hold men's minds together in a common interpretation of reality, there's no hope whatever of anything but an accidental and transitory alleviation to any of our world troubles."[18] You can see Otlet nodding his head in vigorous agreement.

The major difference between the two men involved theory versus practice. Wells was an essayist and polemicist; Otlet tried to build the institutions that would bring his vision to fruition. Wells was playing off his celebrity status. In addition to his utopian novels, *The War of the Worlds* and *The Time Machine* were huge best sellers. (He was also notorious for the conduct of his private life, having had an affair with Margaret Sanger and an illegitimate son with Rebecca West, who was more than a quarter-century younger than he.) But it was primarily his science fiction work that commanded a large general audience. As one of the first hugely successful practitioners he saw himself as an idea man, generating thoughts that others would bring to fruition.

In *World Brain,* Wells expounded on his vision of a great lattice of knowledge fanning out across the globe "like a nervous network"—yet another allusion to humanity's "compromised nervous system." This was more than just a bibliographical challenge, but a project with deep political implications. Technology had started "the abolition of distance," "making novel political and economic arrangements more and more imperative if the populations of the earth are not to grind against each other to their mutual destruction." Wells, like Otlet, had been a believer in the League of Nations and, like Otlet, had come to see it as failing to live up to its promise. By 1938, its failures were becoming clearer by the day. "We are beginning to apprehend something of the full complexity and vastness of the situation that faces mankind," he wrote. "We have to make a new world for ourselves or we shall suffer and perish amidst the downfall of the decaying old." For a model, we needed to turn to the United States, which had become powerful because the railway and the telegraph had replaced the covered wagon and "knitted" it together. "The United States could spread gigantically and keep a common mind." In the same way, he believed the world could follow that example and coalesce into what he called "a world-wide network being woven between all men about the earth."[19]

Whether Otlet's "réseau mondial" or Wells's "world-wide network," the notion was to bring humanity closer together by sharing information and knowledge. "Age by age the World's Knowledge Apparatus has grown up. Unpremeditated. Without a plan," he wrote, and "our World Knowledge Apparatus is not up to our necessities." The World Brain that he proposed would make knowledge freely available through a tightly managed global network that he characterized as "a permanent organism" that would foster cooperation among the world's universities, research institutions, and other centers of intellectual life. The looming Dark Age that Wells—and he was not alone—foresaw demanded nothing less than a radical reimagining of the life of the mind.

To both Otlet and Wells, encyclopedias—the central point of the global networks—were more than just repositories or knowledge, but "a sort of mental clearing house for the mind, a depot where knowledge and ideas are received, sorted, summarized, digested, clarified and compared.... It played the rôle of a cerebral cortex to these essential ganglia," gradually broadening its reach to "extend its informing tentacles to every intelligent individual in the community—the new world community." Working in close concert with universities and other institutions, it would rely on a staff of thousands of specialists to curate and administer the collection. These specialists could be distributed all over the world. "Encyclopaedic organization need not be concentrated now in one place," he wrote. "It might have the form of a network."[20]

Wells also believed that each user of the system would require some form of secure identification—such as a fingerprint—so that he or she "can promptly and certainly be recognized."[21] His recognition of the need for identity management seems foresighted. Indeed, the lack of universal user authentication remains one of the major weaknesses in today's World Wide Web. Today's users must maintain a confusing panoply of user IDs, passwords, and other authenticating details to access information stored with third parties. The resulting operational inefficiencies and security risks are evidenced in the epidemic of hacking and data breaches on the modern Internet. Yet in 1905—years before he had even fully articulated his vision of a world brain—Wells had pinpointed the need for a reliable mechanism to track the personal identity of everyone on the planet. Setting aside the Orwellian implications of such an idea—the notion of governments and their National Security Organizations tracking us all—the wisdom of such an approach within the domain of information access seems evident.

Like Dewey, Geddes, Andersen, and Le Corbusier, Wells was a fellow traveler. And like Otlet, he recognized the enormous challenge presented by "the clamour of statement, misstatement and counter-statement" that characterized so much of the written record and argued passionately that the world needed a more controlled scheme to impose useful order on the "great caterwaul of published voices." Rayward has noted the problematic aspects of Wells's World Brain. For one thing, he seems to proceed from the assumption that the truth is ultimately fixed and knowable—a distinctly Aristotelian supposition that would doubtless strike many contemporary readers as quaint and outdated as a Victorian settee. For Wells, writing during the ascendancy of the British Empire, the notion that learned people might obtain something approaching universal truth may have seemed like a perfectly reasonable supposition. To that end, he imagined that a vast team of thousands of professional encyclopedists would ultimately do the work of sorting fact from fiction, uniting the world's published output into an authoritative body of knowledge. There would be no dissent. Wells's vision was perhaps less authoritarian than collectivist, but the authoritarian aspects of his vision seem deeply troubling today.[22]

The same case could be made of Otlet, whose notion of centralizing authority invited concentration and therefore abuse of power, however enlightened the intentions behind it. James's skepticism of Andersen's World City visions stemmed from his sense of the hopeless naïveté behind it. What Andersen was proposing was not how the world worked. Cities could not be made; they grew organically and by processes that were beyond individual manipulation. He might also have argued that the megalomania of Andersen's scheme invited abuse and control. All dreams of unification are vast simplifications, and contained squarely within the utopian is the dystopian, as Orwell and Aldous Huxley and others were aware, particularly as the Soviet experiment turned out, in Stalin's hands, to be nearly sociopathic.

The dreams of encyclopedists like Otlet and Wells were enlightened ones, an expansion of knowledge that would precisely keep it from falling into the hands of a few (in the way that atomic experiments in the next few years would highlight, and in which Vannevar Bush would be directly involved). And the notion of organizing chaos was predominant. There was, as Wells argued and Otlet agreed, a need for a "greater mental superstructure." A network was not a net, and ultimately Wells, like Otlet, envisioned a system that would include not just libraries but a much larger swath of culture, including museums, galleries, atlas makers, and survey designers. The encyclopedia would consist of carefully edited excerpts and summaries, synthesizing information from multiple sources and presenting it as clearly as possible. Wells himself had made his own attempts along these lines, creating a series of synthetic works on history, biology, and economics. These efforts convinced him of the greater possibilities that might be achieved if he could marshal a small scholarly army of thousands of encyclopedists to create the grand synthesis that he believed the world so badly needed.

Wells's writing circulated widely in his day and profoundly influenced the general public. Otlet incorporated Wells' ideas into his own conception of the "collective brain." In the years following the 1937 conference Otlet continued to make explicit references to Wells, drawing direct parallels between their conceptions of the World Brain. In tribute to Wells, he created his own drawing of Wells's "sick man" metaphor to illustrate the broken state of the world's knowledge-sharing apparatus. Wells was also familiar with the work of another conference attendee, Watson Davis, the American information scientist who created an early microfilm-based syndication service for scientific articles, an endeavor that he hoped might pave the way for a constantly updated "World Bibliography of Scientific Literature" to be distributed on microfilm; he later established the professional association that became the American Society for Information Science and Technology.

There is no clear evidence that Wells's writing influenced Vannevar Bush, who was working on his own ideas for an information storage and retrieval device that he ultimately dubbed the Memex—widely cited as the direct conceptual forebear of the World Wide Web. Scholar Michael Buckland has observed that Bush was notoriously stingy when it came to citing others' work: his landmark essay "As We May Think" contains a notable lack of citations to the work of any of the well-known people who had previously voiced similar ideas: Ostwald, Otlet, Davis, or James Bryce at IBM, who had received several patents for notably similar concepts.[23] Connecting Otlet, Wells, Goldberg, and Bush offers tempting lessons, as well as trajectory that takes technology from being put in the service of peace to that of war, particularly given Bush's activities during World War II and the uses to which Goldberg's invention were put. It involves the connection that James saw between the organic and the mechanical. Creating a brain meant creating a way of controlling its activities and functions, and Otlet's "mechanical collective brain" falls into the evolution from the organic to the mechanistic.

The notion of a "mechanized brain" was much in the air in the late 1930s. That 1937 World Congress on Universal Documentation also introduced Wells to Watson Davis, founder of the American Documentation Institute and a pioneer in science information, and two British documentalists, A. F. C. Pollard and Dr. S. C. Bradford, who were exploring the possibilities of microfilm. Indeed, Watson appears to be the connector linking Otlet, Bush, Wells, and Goldberg. He knew them all,[24] and their thinking influenced his own efforts at promoting the use of microfilm as a transformative information technology.

Just as it had for Otlet, microfilm technology seemed to captivate Wells, who started to explore its potential for effecting a radical transformation in information storage and retrieval. Microfilm offered the prospect of "a real intellectual unification of our race," he wrote. While microfilm itself would never quite live up to that expectation, its successor—digital storage—may yet come close to realizing Wells's vision of a unified "new all-human cerebellum."[25]

In Wells's and Otlet's imaginations, encyclopedists and librarians were to help mankind evolve, putting themselves at the crossroads between the organic and mechanistic notions of the human brain. Technology and evolution were intertwined. Otlet's sentiments about the potential of emerging technologies found another powerful if mystical voice in the person of the French Jesuit priest Teilhard de Chardin. Trained as a natural scientist at the Sorbonne, de Chardin developed a particular fascination with the question of evolution after reading Henri Bergson's popular book, *The Creative Evolution* (*L'Évolution Créatrice*), a Neo-Darwinian treatise that proposed the existence of a universal "vital force" underlying the process of planetary evolution. De Chardin became fascinated with the question of whether human culture evolved according to the same mechanisms that governed biological evolution and began to speculate about the possibility of a coming "Omega Point," a moment of universal enlightenment in which the universe itself attains a level of supreme, interconnected "awareness."

In 1923, de Chardin wrote about his vision of a "noosphere," a kind of ethereal network of human consciousness that, he felt, was drawing humanity ever closer to that evolutionary tipping point. "A wind of revolt is passing through our minds," he wrote, "one that draws us all by a sort of living affinity towards the splendid realization of some foreseen unity." De Chardin saw this as more than an epistemological evolution, but as part and parcel of larger changes in the social, cultural, and political spheres that ultimately would inspire humanity toward a new age of connectedness, transcending old national boundaries and modes of thought. Although he did not share Otlet's positivist outlook—and, indeed, challenged the presumptions of scientific positivism—de Chardin nonetheless recognized some of the same historical forces at play. On the one hand, he felt that modern human beings would become more individuated and, at the same time, paradoxically, more connected. He called this process "complexification": a phenomenon that would ultimately lead to a great convergence of humanity's collective intellectual faculties. He wrote in hopeful terms about the trend toward internationalism and the tendency toward "greater cohesion, justice and brotherhood."[26]

At roughly the same moment an ocean away, a young Argentine librarian named Jorge Luis Borges also began to contemplate the questions of how information might be organized—or not. In 1938, he wrote an essay entitled "The Analytical Language of John Wilkins." Here Borges took satirical aim at the seventeenth-century English bishop's famous universal language scheme, offering in its place a modest taxonomic proposal of his own. He conjured a fictional Chinese encyclopedia called *The Celestial Emporium of Benevolent Knowledge,* which purported to classify all known animal life as follows:

a. belonging to the Emperor
b. embalmed
c. trained
d. sloppy
e. sirens
f. fabulous
g. stray dogs
h. included in this classification
i. trembling like crazy
j. innumerable
k. drawn with a very fine camelhair brush
l. et cetera
m. just broke the vase
n. from a distance look like flies

Borges's essay also took a thinly veiled swipe at Otlet and Henri La Fontaine. Borges, who had recently started working as a librarian in his native Argentina, saw in Otlet a kindred spirit to Wilkins. "I have registered the arbitrarities of Wilkins," he wrote in the essay, "and of the Bibliographic Institute of Brussels."[27]

While there is no evidence of any direct contact between Borges and Otlet, nonetheless Borges clearly knew of Otlet's work. The Universal Decimal Classification had found an early following among South American libraries, and Otlet's name may well have lingered in the air during Borges's time. Why did Borges reserve such particular ridicule for the IIB's effort to organize the world's knowledge? After all, he too had chosen to spend much of his professional life working in the library. As a devoted modernist, however, he believed in the principles of relativism rather than centralized authority. As a self-described "Spencerian anarchist"—a term derived from his admiration for the free-thinking English philosopher who, like Otlet, had been drawn to the positivist philosophy of August Comte but later rejected its idealism in favor of a more relativist, Lamarckian view of the world—Borges felt an innate suspicion of what he considered overreaching utopian schemes, such as the one that seemed to underlie the work of the Belgian bibliographers. Giving voice to that ethos, he wrote, "It is clear that there is no classification of the Universe not being arbitrary and full of conjectures." "The reason for this is very simple: we do not know what thing the universe is."[28]

Like Henry James, Borges was more interested in the personal, the intimate, and the relative than in the sweeping schemes that consumed men like Otlet and Andersen. Both Borges and James seemed to anticipate the coming era of cultural relativism, and to hold out great contempt for universalist, utopian constructions and centralized power that seemed dangerously overreaching. Had Borges taken a closer look at Otlet's writing, however, he might have discovered more of a kindred spirit than he might have expected. While the ontological authoritarianism of the UDC might make an easy target for satire, Otlet fully recognized the limitations of all classification systems. "All classification is naturally, in part at least, a matter of convention," he wrote. "Objectively there exist only distinct objects or separate ideas. All links which we establish between objects or ideas bear the mark of subjectivity."[29] Umberto Eco has argued that John Wilkins's work anticipates the notion of hypertext, insofar as it proposes a framework for drawing connections between related topics by means of symbolic links. Viewed from this lens, says Eco, "many of the system's contradictions would disappear, and Wilkins could be considered as a pioneer in the idea of a flexible and multiple organization of complex data."[30] So too did Otlet, especially in his later work, express a desire for a more complex, nuanced approach to classification that would balance the top-down imperatives of a central authority with the distributed efforts of local governments, universities, research institutions, and even bookstores.

Otlet was no relativist, however; he continued to believe in the positivist dream of a grand synthesis that would one day unite the branches of human knowledge. While acknowledging the fundamentally contrived nature of such schemes, he nonetheless began outlining a new framework for integrating the rapidly proliferating disciplines. The framework was called the Science of the Book and would rely on drawing a clear separation between form and content, or what he called documentation and knowledge. He recognized "the unity and inter-dependence of individual branches of knowledge" but argued for more discipline in the organization and delineation of documents. He made a rather sweeping—and authoritarian—pronouncement on how knowledge gets produced, arguing that all knowledge is properly developed through scholarly societies, which emerge when a critical mass of researchers organize themselves into an association. These societies in turn begin to produce proceedings or other publications, and eventually larger compendiums, bibliographies, and yearbooks with their accumulated findings. They also hold meetings or congresses, which exist to stimulate discussion and support for the collective scholarly undertaking. He talked about the importance of teaching, of organizing the pedagogical enterprise of professorships, courses, lectures, and research institutes.

Ultimately, however, all this effort culminates in the production of untold thousands of documents, primarily books and journal articles. This is where Otlet focused his attention: on the proper editing, production, preservation, and indexing of these works. After all the utopian schemes of World Cities and mechanistic brains, it came back to the printed word and how to organize it. Like Dewey, Otlet believed in the liberating potential of standardization—not machination. If it were possible to standardize the means of scholarly production, knowledge would flow freely.

10

The Radiated Library

One day, Paul Otlet's grandson, Jean, was taking a walk on the beach with his grandfather when they encountered some jellyfish washed up on the shore. Many years later, he shared his recollections of that day with Otlet's biographer, Françoise Levie. Jean remembered watching as his grandfather stooped down to pluck the jellyfish out of the sand, one by one, and stack them on top of each other. He then reached into his jacket pocket and produced an index card, on which he proceeded to write down the number "59.33": the Universal Decimal Classification code for Coelenterata.[1] He placed the card gently atop the pile of soggy invertebrates, and walked away.[2]

As Otlet entered his later years, he started to withdraw from the outside world. Or perhaps it would be more accurate to say the outside world withdrew from him. After the Belgian government shut down the Palais Mondial once and for all in 1934, younger men like Donker Duyvis had taken the reins of the IIB, and Otlet found himself increasingly isolated. He had little choice but to accept his waning influence over the affairs of the country he loved and the institutions he had founded. Settling into the role of *eminence gris*, he removed himself from the bureaucratic squabbles of the IIB and repositioned himself as an organizational figurehead. Although he tried to remain active—writing opinion columns for the papers and corresponding with old friends—he increasingly turned his attentions inward, retreating back into the intellectual cocoon that had enveloped him in his youth. He began to reflect and refine his conceptions of the ideals he had championed over the years: universal knowledge, world peace, and, most mystically, humanity's potential for transcendent realization.

As he started to pull back from the public sphere, Otlet began to channel his energies into his writing. He penned articles on a vast range of topics, including "World Law: International Law," "Universal Rights," "'Cosmo-Meta-Political Law,'" and the even loftier-sounding "The Future and Infinity, Eternity and God."[3] Few were actually published. He seemed to be working out his ideas for a larger work that would encompass them all. In 1934 and 1935, however, he published the three books that would summarize his intellectual legacy: *The Treatise on Documentation* (*Traité de documentation*), the succinctly titled *World* (*Monde*), and *The Plan for Belgium* (*Plan Belgique*).

The energy and drive on display in these books stand in stark contrast to the sedentary, inwardly focused life that Otlet had started to lead. The most exhaustive of these works, *The Treatise on Documentation,* makes for heavy going at first: 411 pages of densely set, two-column type, it seems at first blush to resemble a technical manual more than an encyclopedist's manifesto. With numerous sections and subsections carefully demarcated and enumerated, peppered liberally with quotations and footnotes from other works, the book has a stitched-together character that at times feels like Exhibit A for Otlet's monographic principle. Easy reading it is not. Rayward describes the book "as a kind of *reductio ad absurdum* of Otlet's thought."[4] But for all its density and sometimes plodding repetition, the book captures the full sweep of Otlet's vision, and in a few passages toward the end it delivers stunningly prophetic glimpses of the twentieth-century information age then coming into view.

"The huge mass of published material grows by the day, by the hour, in amounts that are disconcerting and sometimes maddening," he wrote. "Like water falling from the sky, it can either cause flooding or beneficial irrigation."[5] Pondering the peril and the promise of this growing data deluge, Otlet brought his positivist philosophy to bear, framing the predicament in terms of a larger historical arc.

Otlet recognized that the concept of universal knowledge has had a long history—the Library at Alexandria, the great medieval encyclopedias, Conrad Gessner's Universal Bibliography, and the later, heroic efforts of nineteenth-century librarians and bibliographers to stem the rising tide of books. He situated his own efforts squarely within that lineage. "The universal bibliography is essential," he wrote, "to offer a complete summation of the great work of humanity, from the earliest times until our current civilization. The sum total of these collections is a true story of the human spirit."[6]

Otlet introduced each major section of the book in terms of a multiphase evolutionary process. At the outset, he argued that human communication had evolved through three distinct eras: in the first, oral culture predominated, as knowledge spread through word of mouth; in the second, writing evolved, and the document took center stage as the premier manifestation of human thought; now, in the third era, machine-based communication would predominate: telegraphs, telephone, postal services, radios, and so on. Books, no longer commanding the dominant cultural position they had enjoyed for centuries, now had to coexist within a wider landscape of periodicals, films, photographs, and other forms of media, all coexisting in what he called a "simultaneous culture."

Similarly, libraries and archives have gone through three phases. In the first phase, libraries acted as centers of learning, building catalogs of available knowledge and fostering the creation of catalogs, compilations, dictionaries, and encyclopedias. In the next phase, bibliography emerged—driven by the needs of scholars to gather and collate resources to suit their purposes. Now, Otlet saw a third phase taking shape, one in which traditional bibliographies gave way to a new model of knowledge production in which books fused with other kinds of documents, transforming them into broader and more flexible knowledge structures.

Museums too were following much the same trajectory. In the first phase, they existed primarily as conservators of rare objects; in the second, they began to house facsimiles (such as reproductions, photographs, and other simulacra), to give a sense of thoroughness; in the third, museums would begin to federate their collections, displaying material across multiple domains and physical locations—the kind of syndicated, widely distributed museum experience that Otlet and Neurath had envisioned a few years earlier. Geographical limitations would no longer constrain the scope of their curations. Ultimately, all of these institutions—museums, libraries, bookstores, schools, and other research facilities—would coalesce into a unified network for intellectual exchange.[7]

Vision though this was, Otlet pursued it all in excruciating detail, chronicling the evolution of recorded knowledge over the years—from the origins of writing to the latest communication technologies, such as television, radio, and microfilm. The book chronicled the history of signs and symbols, alphabets, writing instruments, maps, the economics of paper-making, and a history of printing, with extended detours into the arts of stenography, cryptography, Braille, Morse code, and even the so-called spirit-writing of mediums (what present-day New Age acolytes would call "channeling"). He then went on to catalog all the many possible permutations of writing through the ages: philosophical, scientific, mathematical, personal, poetic, novelistic ("a detestable genre," he wrote, half-mockingly, "because it does not permit drawing conclusions"[8]), biographical, encyclopedic, journalistic, dissertational, and what he called "enigmas," like the Egyptian Sphinx.

This comprehensive cataloging of literary forms led Otlet to consider the effects of technology on the production of knowledge, exploring how successive waves of innovation influenced the diffusion of human thought. Movable type, steam-powered printing, photo engraving, rotary presses, and mechanical composing machines had all contributed to the rise of the publishing industry, and eventually the emergence of modern newspapers and other popular literature.[9] He traced how these tools had helped to fuel the spread of democracy and constitutional systems based on freedom of the press, and how the emerging technologies of transportation, telegraph, and telephones were now giving rise to a broader spectrum of recorded thought: radio, moving pictures, phonographs, lectures, theater performances, sculpture, architecture, and other so-called spectacles. All together, these diversifying forms of recorded knowledge would comprise what he called "substitutes for the book."[10]

Otlet took a particular interest in the possibilities of cinema, cataloging the differences between various film formats and waxing philosophic on the ability of cinema to "carry the seed of unanimous expression, sincere and exclusive to the modern world." Comparing the expressive power of motion pictures to Greek tragedy and medieval cathedrals, he imagined films escaping the confines of the rectangular projection screen, imagining a futuristic environment in which films might be projected onto the interior of a sphere with multiple images juxtaposed around it—a notion not far removed from the contemporary IMAX cinema display. He also wrote, again at length, about the possibilities of the still-nascent technology of television. Together, this growing panoply of new media forms would converge into something new, a higher level of expression that would one day allow humanity to approach the divine, to achieve "the radiant contemplation of Total Reality."[11]

For his entire life, Otlet had worked to extend definitions, to move the boundaries of expression and understanding. In his earlier writing, he had used an expansive definition of the term "book"—useful though it was—to encompass all the varied forms of recorded knowledge. Now, he proposed a neologism: the Biblion. Physicists had already discovered the power of the atom (although they had not yet figured out how to split it). Similarly, Otlet claimed to have discovered the secret to breaking down recorded knowledge into its component units, then re-assembling those units into new forms—Biblions.[12] Unlocking the potential energy of these building blocks would require a new kind of science: "bibliology." As opposed to bibliography—which merely tries to describe the contents of books—bibliology aspires to a higher purpose: guiding the production and distribution of all kinds of recorded knowledge.[13] The bibliologist would do more than simply gather information; he or she would enable access to every domain of human knowledge, create new works, and "provoke inventions" to help support the spread of human knowledge.

By following a rigidly controlled set of cataloging rules—based on the Universal Decimal Classification—the bibliologist could synthesize material from multiple sources, thus allowing others to search, retrieve, and repurpose the contents in theoretically endless permutations. The cataloging system—relatively unchanged from when Otlet first conceived it half a century before—would provide a framework for describing every fact, element, illustration, photograph, diagram, and so forth contained in each source, providing rules for everything from subject classification to the proper treatment of titles, pagination, and the notation of divisions in a text. Here is Otlet, the master cataloger, at the top of his game. He lays

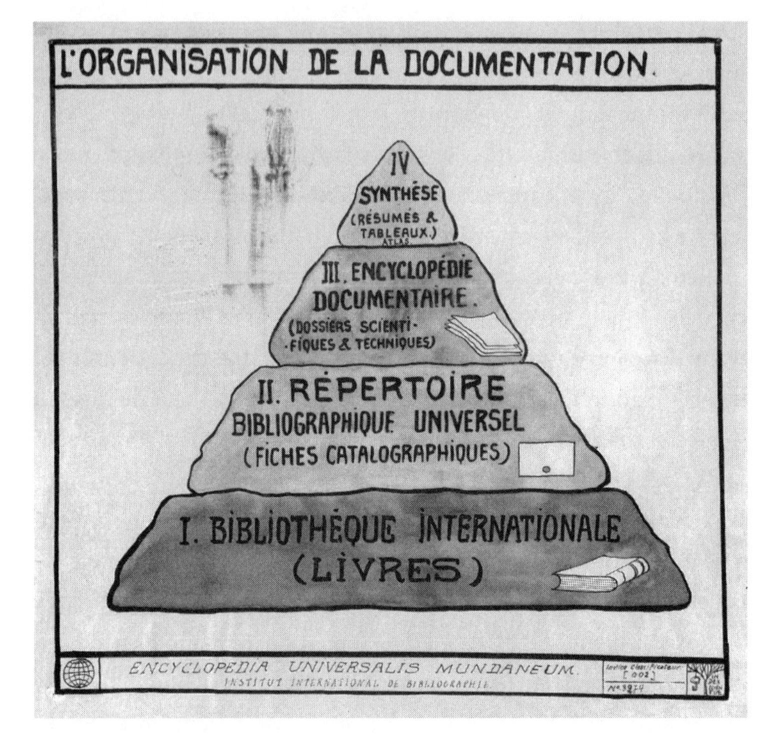

The Organization of Documents (*L'Organisation de la documentation*). Unpublished drawing from the Encyclopedia Universalis Mundaneum papers. Reproduced with permission from the Mundaneum.

out exacting rules for citing every possible permutation of an author's name (including authors writing under multiple names, pseudonyms, or even anagrams) and how to cite page numbers, volumes of serial publications, and all manner of other bibliographic arcana. Finally, he describes how each atomic unit of information would receive a fixed locator record to ensure its availability to the larger population of users.

And lest anyone think that Otlet had given up hope for the grandiose transnational institutions he had once proposed—culminating in his collaboration with Le Corbusier—here he argued that this whole biblionic enterprise could only succeed through the aegis of a central coordinating institution, a Universal Organization, working under the auspices of a broadly empowered international body such as the League of Nations, which, though politically toothless, might yet exert an influence over the world of knowledge. What followed was a complicated hierarchical schema of national-international cooperation that would allow the UDC to encompass all the world's information in a Universal Network of Documentation, facilitating the distribution of books, newspaper and magazine articles, audio recordings, films, and photographs all over the world.

For all his bureaucratic zeal, Otlet also realized that by themselves new organizational hierarchies would never bring this vision to life. It would require nothing less than remaking the entire economic chain of knowledge production. "What do writers receive in return for their work?" he asked. "This question has arisen in every age, and particularly in our own because of three factors: the increased number of intellectuals, the increasing role of intelligence in society, and new forms driven by commercialization and professionalization." Whereas authors in antiquity and the Middle Ages may have written for their own pleasure or for the glory of God, the economic realities of the industrial age had changed that equation. "One uses pen and thought like bread and wine," he wrote. "Our age has turned literature into a trade."[14]

Still, Otlet held out hope for what he called the "disinterested task" of the writer, ultimately envisioning something akin to a gift economy of intellectual work (a notion that has attracted considerable currency among contemporary Internet pundits). He hoped—with perhaps a characteristic surfeit of optimism—that enlightened knowledge workers would one day see the benefits of embracing the monographic principle in their own work, using index cards for note taking, and making their finished work more accessible by breaking it into index-able chunks of information that could move more fluidly through the global knowledge network.

Still, he recognized that knowledge workers needed to be paid, at least in the here and now. In considering the financial implications of his scheme, Otlet saw a special role for bookstores, given their role as the primary gateways to knowledge for many readers (public libraries were far less common during his era). He also imagined that a central catalog based on the Universal Decimal Classification could be made available to booksellers, so that any given bookstore could theoretically provide access to any book ever published.[15] Each bookstore would then function as a node in the international network—this *réseau mondial*—so that the catalog doubled as both a bibliography and a virtual inventory of available works.

So much was changing so fast, and yet the true transformation had not yet happened. Human knowledge remained largely stuck in books, such as the very one he was writing. "Despite advances in scientific thought and bibliographical material, the modes of recording knowledge have not made much progress."[16] But it would happen. Toward the end of the Treatise, Otlet revealed his vision for the future, one that he described variously as a machine, as an organism, and, ultimately, as a vehicle for spiritual transcendence.

First, the machine. Otlet cast the book (again, using the term in the broadest possible sense) as a "book-machine," a node in a larger apparatus that would, over time, beget a geometric expansion of human knowledge.[17] The book-machine, he argued, functions like a "mechanism" or "dynamism," loaded with potential energy. That energy travels through a vast network (*réseau*), exposing the implicit associations of ideas flowing both within and between other book-machines. "Right beneath our eyes, a great machine for intellectual work is coming into view," he wrote. To describe how this machine might work, he turned to other familiar reference points: railways, post offices, telegram, telephone, radio, and other industrial networks that were now shaping the flow of information. One day, he imagined, a global network structured along similar lines might allow professors to broadcast their lectures to far-flung audiences, perhaps even allowing for the audience to interact by asking questions over the telephone line. Similarly, international associations might be able to convene their members from a distance, using audio and video (what we would call "virtual meetings").[18] The network would require a high degree of standardization in terms of protocols and administrative processes. Just as railways, post offices, and electric utilities had benefited from adopting universal standards, so the global network (*réseau mondial*) would require a widespread commitment to technological standards.

The machine metaphors stem from the concept of the book as a unit of stored energy, containing condensed thoughts that could then expand in the brain. The book-machine functioned within the larger system as both an "accumulator"—an externalized repository of knowledge—and as a "transformer"—an agent for producing new knowledge within the larger ecosystem.[19] As these little machines propagate, following the patterns of repetition and amplification that have always shaped the spread of knowledge, they collectively start to comprise a larger machine. In this way, the entirety of human civilization could be construed as a kind of vast knowledge machine: The Mundaneum.[20]

Within that universal knowledge machine there would be many submachines, each providing an open-ended and never-finished synthesis of all the available information on a particular topic. Ultimately, these collections might be condensed into storage units of minute proportions, such that each was part of the whole, and the whole connected to each part—an idea that in many ways anticipates the networked personal computer. Otlet's list of technical features included a portable writing machine that could fit inside a briefcase, a microphone for dictation, a calculator, and a "selection machine" for retrieving information stored on punch cards (he refers to the work of Herman Hollerith, the punch-card inventor whose company embraced the card-cataloging technologies of Melvil Dewey's Library Bureau, and eventually morphed into IBM). Over time, he saw these functions converging into a single device capable of carrying out most of the functions of the personal computer: reading, writing, viewing photographs, browsing archives, attending "telelectures" at a distance via a television screen, recognizing and transcribing speech, and even automatically translating documents between languages. "Today these are separate machines," he writes. "Grouping them together in a single unit will amplify their performance."[21] Looking even farther ahead, he contemplated the possibilities of incorporating television, sound recordings, and even a kind of tactile documentation that could convey the senses of taste or smell. Otlet was inspired by the machine's capabilities. He wrote, "One could imagine an electric telescope able to request pages from books in the great libraries and project them in a telegraph room."[22] His vision of what went on in the Mundaneum deserves full citation, for its Wells-like, even Jules Verne–like, scope:

> Here, the desk no longer carries any books. Instead there is a screen connected to a telephone. Over there, in a great building, are all the books and related material, with all the space necessary for cataloging and registering them, with all the catalogs, bibliographies and indexes.... From there one could call up a page on screen to read the answer to questions posed by telephone. The screen could be double, quadruple or [decuple] if

> there are multiple texts to show simultaneously; there would be an audio speaker if needed for additional material to complement the text. Wells certainly would have liked this idea. Today, this seems like utopia because there has never existed such a thing, but there could come a day when our efforts could lead to this perfection.[23]

In later writings Otlet would describe a device called a Mondotheque, which looked something like a contemporary computer workstation. It was a desklike device equipped with a collection of electronic instruments: a radio, telephone, microfilm reader, television, and record player, as well as a collection of personalized documents that might consist of selected books, movies, photographs, and so forth. Each workstation would include copies of the Universal Decimal Classification; the Atlas Mundaneum, a volume that attempted to synthesize the highest levels of scientific knowledge; the Sphaera Mundaneum, a kind of visual index that would allow users to pore over information graphics intended to provide high-level overviews of certain domains of knowledge; and the Pyramid Mundaneum that presents the contents of the collection in a hierarchical, nested grouping of categories. It would be an autonomous machine that allowed the user to create a highly personalized information environment arrayed around one's personal interests but directly connected to the larger corpus of recorded knowledge. The Mondotheque was more than just a platform for consuming information; it was an active tool for knowledge production. It would include a large desk space for writing and drawing. Most important, it would remain tethered to the Universal Network of Documentation via a persistent connection.

To function effectively as part of the global network, Otlet suggested, it would have to satisfy a number of core requirements:

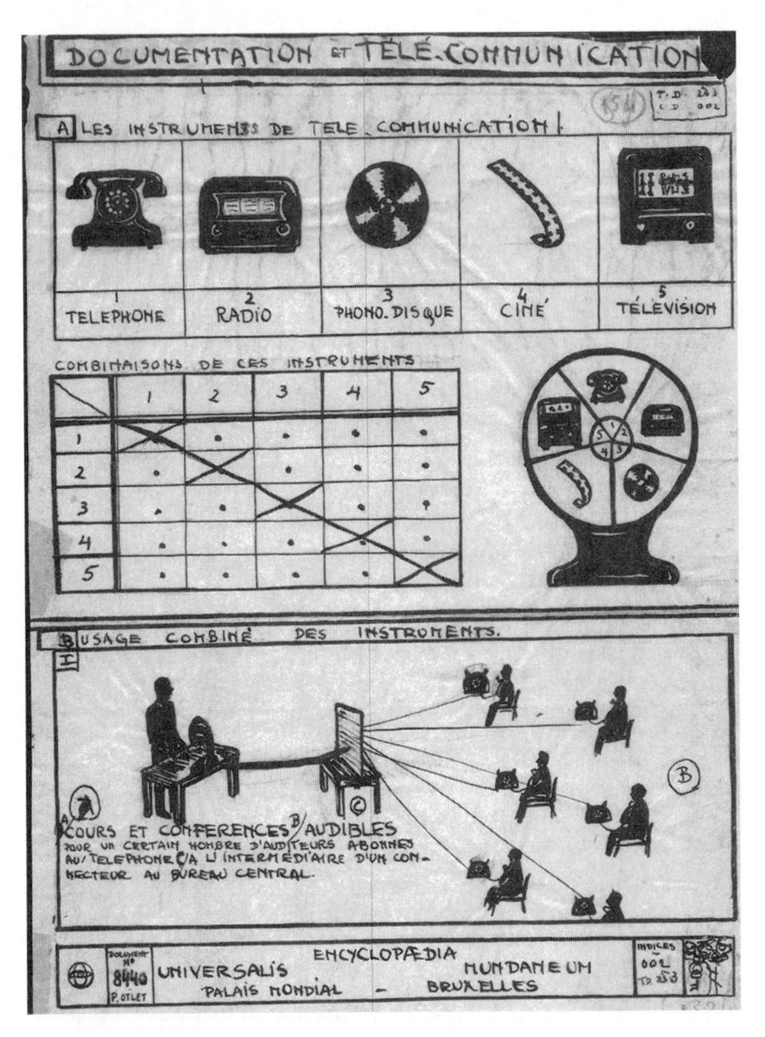

Unpublished drawing from the Encyclopaedia Universalis Mundaneum papers, 1930s. Republished by permission of the Mundaneum.

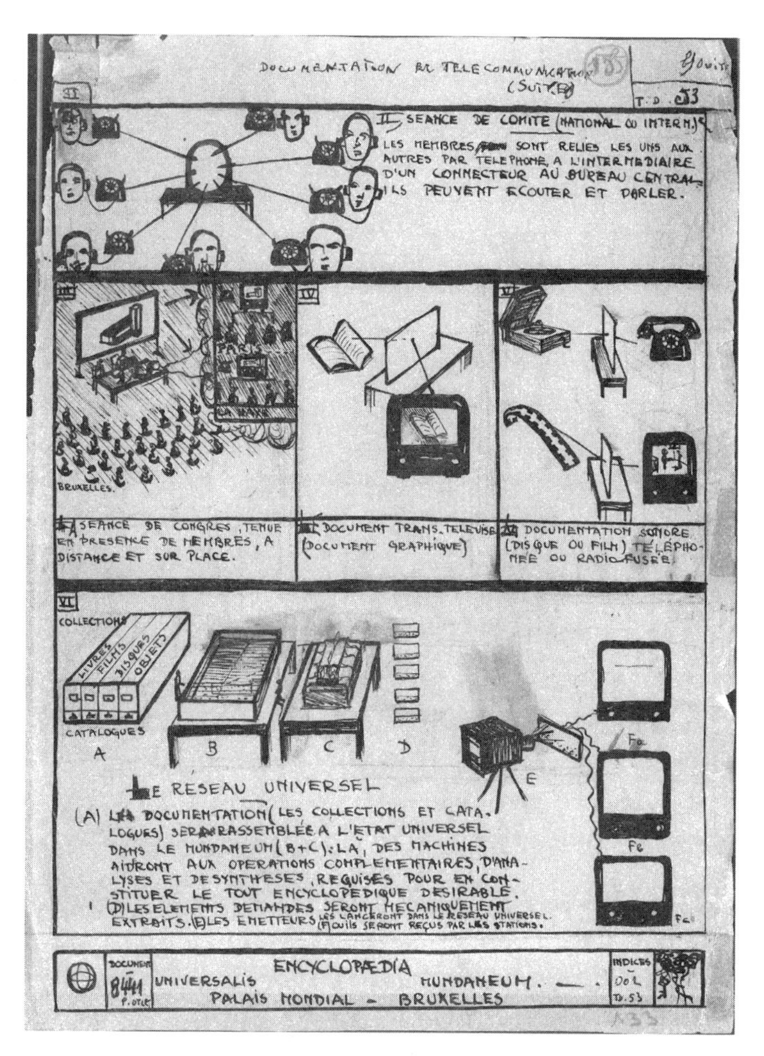

Unpublished drawing from the Encyclopaedia Universalis Mundaneum papers, 1930s. Republished by permission of the Mundaneum.

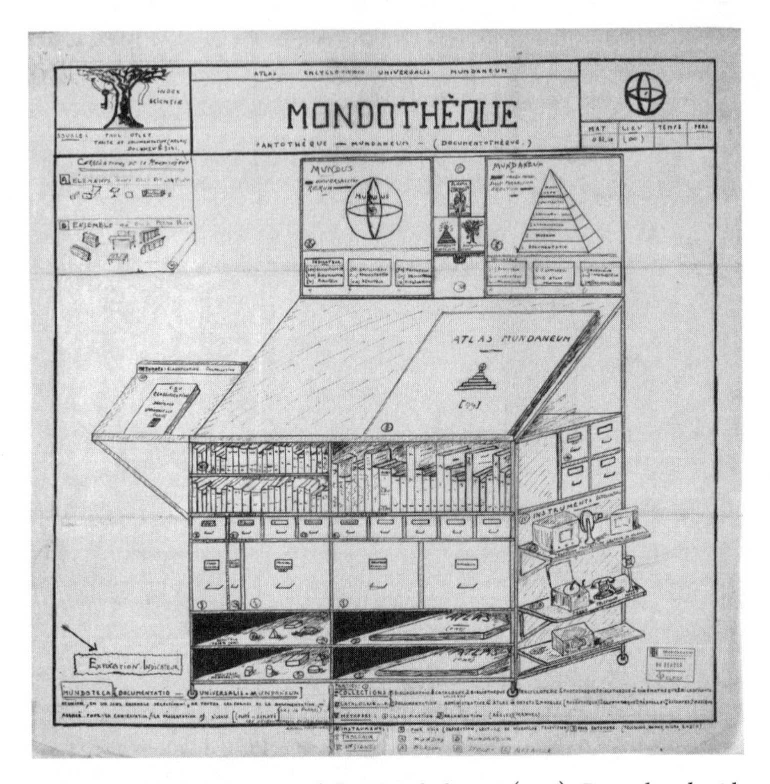

Otlet, unpublished drawing of the Mondotheque (1941). Reproduced with permission from the Mundaneum.

reproduce copies of a single text over and over again; display the relationships between documents through explicit links; provide support for a classification system to govern those associations; retrieve stored documents automatically; and recognize human speech (a facility that he described "as the ability to transform sound into writing"). If inventors could perfect the machinery required to realize this vision, they would then have created a "mechanical and collective brain."[24]

Such were the mechanical aspects of the Mundaneum. But it was to be more than just a collection of tools and processes. Otlet also envisioned it as a living organism: a "body formed from a collection of organs" that would adapt over time to its surroundings, eventually coalescing into a kind of collective sentient being. "All the phenomena of life form a kind of network, in which nothing moves without causing oscillations elsewhere. So it is with the elements of a book, and with books between each other."[25] The growth of the Mundaneum would follow the same evolutionary course. Human thought was itself a kind of "organism," following the same laws of philogeny, ontogeny, and heredity that governed organic life. "The network is organically and hierarchically constituted," he wrote, "in such a manner that local agencies will be linked to regional ones, these to national ones, the latter to the international and the latter to the world organism."[26] That distributed, networked consciousness—echoing Wells's notion of a "new all-human cerebellum"—was the essence of the Mundaneum.

Finally, the Mundaneum existed as a pure ideal, a psychic and spiritual force for change. "The law of change and evolution is universal," he wrote. "Everything in nature is in a state of perpetual change, always in the process of becoming."[27] The same natural forces that shaped the physical world, the animal realm, and the development of human society also governed the spread of ideas—and opened the door to spiritual realization. It was all part and parcel of the evolutionary process.

In his positivist search for universal laws, Otlet identified the central dynamics that seemed to govern all of these phenomena: adaptation, repetition, and opposition. Of these, repetition played a key role in the dissemination of ideas, through a process he called "amplified repetition." Knowledge spreads through repetition from speaker to listener, or from writer to reader. The more repetition, the more widely that knowledge gets disseminated through the population. But repetition is not enough by itself; knowledge always takes shape in a larger social context and must be adapted and refined over time as individuals evaluate, accept, or reject a piece of information. The process of knowledge production adheres to the same laws that govern the physical transformation of energy. Written knowledge contains a kind of "thought-energy" that is never lost but rather transformed. "Books conserve mental energy," he wrote, and "their contents shift into other books when they are destroyed; and all literary creation, no matter how original and powerful, involves the redistribution, combination, and new amalgamations of existing facts."[28]

So the Mundaneum reflects, in the end, an expansive redefinition of books, which, after all, were the technology that he had at hand. He might dream of other forms of communication, electronic or mechanical, but the written word remained a fundamental unit of intellectual energy. Books observe natural universal cycles as written material is gathered and published, then finds its way into other forms of publication. Not only did he see books as guiding the evolution of the physical human brain, but he also believed that the development of human thought itself followed an evolutionary trajectory.

Hand tools had helped early humans evolve and progress. Books too were essential to humanity's intellectual evolution. "The book is to the brain as the tool is to the hand."[29] This co-evolution of human intellect and its tools of thought reflected what psychologists today might call embodied cognition. As individuals shared knowledge with each other through speech, writing, and other forms of externalized expression, they developed what Otlet characterized as a kind of collective intelligence—a faculty that allowed groups of people to solve large-scale social problems by working together in networks.

"Now a new era has arrived," he wrote, "that must serve the divine purpose of individual well-being." Otlet saw the present age as an axial point in the development of humanity. Human history consisted of stages that had led to the point where such things as a Mundaneum were possible. In the first stage of existence, human beings perceived reality with their senses alone; in the second, they interpreted their experience and gave it externalized expression; in the third, they introduced writing to register their sense perceptions; in the fourth, they created scientific instruments to record empirical data; in the fifth, documents and instruments began to merge into a single unified device—something like the "book-machine" he had written about earlier in the book; in the sixth and final stage, human consciousness itself would merge with these instruments to achieve what Otlet dubbed "hyper-intelligence."

In this enlightened state of symbiosis between humanity and machines, recorded media would merge with the direct recording of human perception—such as sound, taste, and even smell—blending into a state of "hyper-documentation"[30] (a term he coined thirty years before Ted Nelson's more famous neologism of "hypertext"). In a passage that would make the editors of *Wired* blush, he described the possibility of achieving "a pure spirit with access to complete and intuitive knowledge of all things at every moment." Such a state of transcendent realization would not come easily. It would require diligent effort, continual investigation, and ongoing elaboration. In this, it reflected the human condition. "If man is a fallen angel, then he must regain his pristine state through self-improvement and the promise of salvation," he wrote. "If, on the other hand, he is the continuation of a natural line of beings, then he must strive towards higher attainment by finding in his previous ascension the potential for future progress."[31]

The day of ultimate awakening would remain far in the future, however. In the meantime, Otlet imagined the potential steps along that path, proposing three potential hypotheses. The highest-level ideal would entail a state of divine omniscience—similar to the gods or angels—where everyone attains a state of clairvoyance and premonition and renders books completely superfluous. A less transcendent but still transformative state of affairs would involve condensing the entirety of human thought into a compact form that could be placed within arm's reach, like the hypothetical miniaturized *World Encyclopedia* that might serve as a portable appendage to the brain.

What Wells and others had dreamed, Otlet had started to describe in precise mechanical detail. And the machine he described resembles nothing so much as a contemporary Web browser, a machine for retrieving data stored remotely and linked into a vast network. In Otlet's final depiction of the Mundaneum—an unpublished drawing from 1943, a year before his death—he sketched out the contours of what looks like a giant radio tower, beaming out information drawn from documents and transformed into radio and television signals that could be received by end users anywhere in the world. It had become a "thinking machine."[32]

Underlying Otlet's theories and philosophies is the doctrine of universalism, the same cause he had embraced in his youth. A half-century later, his faith in human progress seemed to remain undiminished as he wrote of his conviction in "the fundamental unity of things and the interdependence of everything that comprises them."[33] Cataloging thought would ultimately yield a coherent system of thought and action, of harmony and sensibility, and bring about a state of spiritual transcendence. Like John Wilkins three centuries before him, he believed that it would yield a "universal written language" capable of exerting a powerful influence on the shape of human thought. Through such an ambitious undertaking of what he called "meta-bibliography," he argued, "one could approach the transcendent regions."

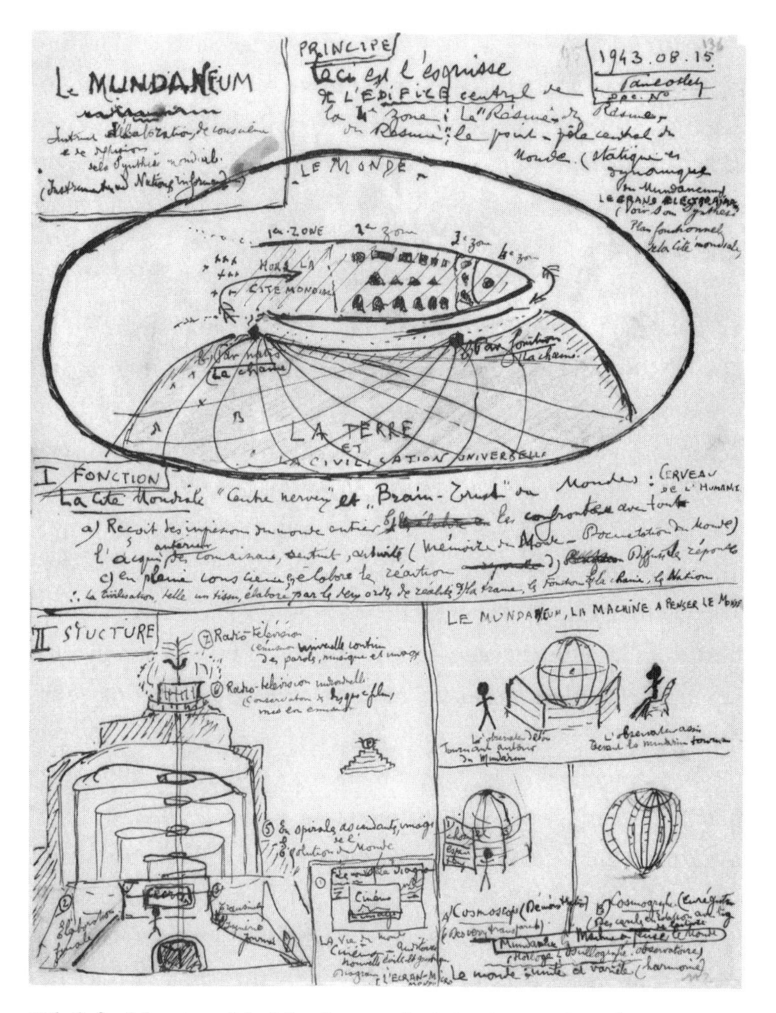

Otlet's final drawing of the Mundaneum, depicting it as a radio-television transmitter radiating information throughout the world (1943). Reproduced with permission from the Mundaneum.

In his follow-up book, *World* (*Monde*), Otlet elaborated further on his notions of universality. It is a more readable book and targeted at a general audience. While it covers much of the same ground as the *Treatise*—including the laborious positivist interpretations of world history—it strikes a more exhortatory tone, placing far more emphasis on the philosophy of universalism and less on the mechanics of cataloging.

Invoking the "extreme interdependence" of the modern world, Otlet discussed the necessity of universalism for social progress and world peace. Although human beings tend to see themselves as independent and self-directed, he argued, in fact their intellectual lives are deeply intertwined with one another. "All substances, all beings and all forms of media act and interact with each other," he wrote. The creator of any new work always "combines and amalgamates existing things to produce new things."[34] And while the act of intellectual creation might seem to spring unaided from the individual mind, in fact all such works are "nothing less than intersecting nodes on a vast social network."[35]

Just as human beings consist of aggregate physical components like atoms and electrons—that combine to create the illusion of a separate, autonomous self—so the creation of human thought takes place in a wider social context. "Societies are *organisms*," he wrote. "An organism is a collection of living cells grouped together, carrying out discrete tasks for the benefit of the whole."[36] Societies are also, he wrote, "unions of intelligence" predicated on autonomous intellects working independently toward the higher good of collective wisdom. To that end, he held out hope for an "intellectual machine" that anyone could use to collect, create, and disseminate information across the wider social sphere. "This instrument constitutes the body of knowledge," he wrote. That body would demand "a new language to express data and intellectual output in documents and other new forms." Such a language would support a new, universal synthesis of human knowledge, as well as new forms of written expression. It would also, he believed, lead to transformations of the collective consciousness itself.[37]

Again he saw this evolution in spiritual terms. Here invoking theosophy—the Buddhist-inspired worldview that had influenced his work with Le Corbusier on the Mundaneum, and whose teachings would hold a particular fascination for the Nazi Rosenberg Commission a few years later—he explored its teachings on the planes of natural phenomena that proceeded from solid to liquid to gaseous, and on to the higher "etheric" realms of consciousness. All phenomena—including humanity—exists in a constant state of flux, of becoming and deteriorating. In the same way, human intelligence is not a fixed thing, he wrote, but rather a kind of "universal fluid" that moves among us and throughout the larger "psychological body of society."[38]

That collective body, like human beings, is susceptible to a misplaced sense of identity. Much of the conflict in society arises from a kind of neurotic grasping for a solid sense of self (like extreme nationalism). The antidote to such strife would involve a new, universal governing structure that recognized humanity's fundamental interdependence.[39] Such an entity—which Otlet still saw as a future version of the League of Nations—would create a new international political system, a monetary policy designed to ensure the fair distribution of wealth, a judicial system, a global language (Esperanto), and foster a great "prostration of spirits" in the service of humanity.[40] As outlandishly ambitious as such a program might sound, Otlet believed fervently that all of this and more might become possible through the widespread application of universalism, which he described as "a system that would embrace all things."[41] Equipped with this worldview, any individual might "perceive that his conscience, his will, his feelings are only aspects of the grand total," he wrote. "Distinctions disappear, and what remains is pure in all its fullness."[42]

In 1935, Otlet further explored the intersection of his universalist thinking with his long-simmering vision for a World City, in a book called *The Plan for Belgium* (*Le Plan Belgique*). Although two decades had passed since his failed collaboration with Andersen, the sculptor's vision had left a deep impression on Otlet, who now called for the inclusion of a World City as part of a sweeping plan for a complete remaking of Belgium. He envisioned the project as "a symbol and visible representation of the idea of the unity of humanity."[43]

That same spirit of collective spiritual realization also marks the end of the *Treatise*, wherein Otlet imagined a future state in which one might "be everywhere, see everything, hear everything and know everything." Is not "this perfection and plenitude that man, in sovereign homage and sovereign good, has attributed to God Himself?" he wondered. "Such, in the end, is the power of the book!"[44]

With these words Otlet brought the *Treatise* to a stirring conclusion. But he also added a coda, in which he bitterly recounted his version of the events surrounding the closure of the World Palace. He accuses the Belgian government of shutting down the World Palace "through an arbitrary act of force," placing this on a par with the failure of the League of Nations and his inability to extract justice from the International Courts. If the World Palace were to remain closed, he laments, "then it seems as though there is no place in our civilization for the ideals indicated in these words over its entrance: Freedom, Equality and Worldwide Brotherhood."

Both the grandiosity and petulance reflect Otlet's disposition during this period, when the world—and not just his own dreams for it—seemed to be collapsing all around him. Belgian self-destruction now seemed unstoppable, and he considered whether to offer the World Palace to some other more welcoming place (he considered the Hague, the United States, and later, in a fit of desperation, Nazi Germany). Curiously and ominously, the book concludes with a cartoon image showing a boot crushing the World Palace. An idealized vision of the Mundaneum stands in the background, shimmering like a beacon, but the foreground holds the immediate future, and it is an image that would prove all too prophetic.

From the *Treatise on Documentation* (*Traité de documentation*), 1934.

11

The Intergalactic Network

Ever since Al Gore's famously misquoted contention during the 2000 presidential election, the question of who invented the Internet has been hotly and endlessly debated. As Gore himself was quick to point out, credit cannot and should not go to any one individual. Tim Berners-Lee, for example, did not invent the Internet. Nor did Vannevar Bush, H. G. Wells, or Paul Otlet. Most wisdom on the subject has now settled on a far-flung group of researchers funded by the U.S. Department of Defense during the Cold War. In response to the Soviet Union's 1958 launching of the *Sputnik I* satellite, President Eisenhower established a scientific-military organization whose goal was to develop strategic technologies: the Advanced Research Projects Agency (ARPA). In 1962, under President Kennedy, that agency established a new computer science–oriented unit called the Information Processing Techniques Office (IPTO), led by MIT professor J. C. R. Licklider.

Licklider had earned his reputation as a thought leader in the still-young computer science world, having written a work called *Man-Computer Symbiosis* and publishing a series of memos for what he jokingly described as an "intergalactic network" of connected computers: perhaps ten or twelve in all, attached to a variety of disk drives, remote consoles, and teletype machines. Ultimately, that network came to consist of a small group of computing centers located at several leading research centers, including MIT, Carnegie-Mellon, the University of Utah, the Rand Corporation, Stanford Research Institute (SRI), and the University of California campuses at Berkeley, Santa Barbara, and Los Angeles. While working with the SRI team, Licklider championed the work of a researcher named Douglas Engelbart, who spent much of the 1960s working on this early proto-hypertext tool known as the oN-Line System (which I will discuss further later).

With the financial backing of the U.S. military, Licklider launched a series of landmark projects that would also pave the way for, among other things, the modern UNIX operating system, which provided a template for solving a larger problem: how to engineer a network that allowed a number of far-flung computers—running on different systems and with different operating capacities—to communicate with each other. "Consider the situation in which several different centers are netted together," as Licklider put it, "each center being highly individualistic and having its own special language and its own special way of doing things. Is it not desirable, or even necessary for all the centers to agree upon some language or, at least, upon some conventions for asking such questions as 'What language do you speak?'"

By establishing a set of standards, Licklider believed he could create a much richer shared-computing environment, one that would allow groups of researchers (or military personnel) in different locations to work together more effectively. Licklider also put "the user"—now a familiar construct in software design circles—front and center in the design of this network. "It seems easiest to approach this matter from the individual user's point of view," he wrote in a 1963 memo to his collaborators at the computing research centers, "to see what he would like to have, what he might like to do, and then to try to figure out how to make a system within which his requirements can be met."[1]

Licklider also happened to have a strong interest in the future of libraries. In 1965, he published a research report in which he described the contours of a global, networked knowledge environment that sought to liberate information from the confines of books. Licklider took a dim view of the traditional bound-book format, which he saw as a terribly inefficient medium for data storage and retrieval. He wrote about "the difficulty of separating the information in books from the pages"—a problem that, he argued, constituted one of "the most serious shortcomings of our present system for interacting with the body of recorded knowledge."[2] What he would do was create a system for cataloging information from a wide range of sources, extracting and indexing that information—even down to the level of individual sentences—and distributing it over a network.

By the 1960s, the notion of extracting written information from books and storing it elsewhere seemed increasingly plausible. The tools had started to emerge, including electric typewriters, photocopiers—using a technique called xerography—and, increasingly, computers. Microfilm had established itself as a viable technology for nearly three decades, and a new generation of office workers in America's postwar boom economy churned out documents in unprecedented numbers. Many of these new technologies, directly or indirectly, were attributable to the U.S. Department of Defense, which devoted an increasingly large portion of its budget to developing cutting-edge computer technology and networks.

Innovative thinking about electronic information was percolating outside the military world as well. In 1951, the American librarian and information-technology specialist Jesse Shera suggested that the contents of books might be liberated from traditional formats. "Classification, then, can achieve its fullest purpose," he wrote in a 1951 conference paper, "only after the *idea* content of the book has been dissociated from its physical embodiment—its codex form."[3] Shera saw a central role for the skilled knowledge worker, using sophisticated mechanical tools to collect, collate, and synthesize material from an ever-growing influx of data. "The librarian of the future," he wrote in a 1957 paper, "may well be regarded as the geneticist of our intellectual life."[4] Shera never shared Licklider's level of concern with the end user, however. He tended to focus more on an idealized social outcome, rather than on the felt experience of the individual human actor.

For his part, and supportive of libraries though he was, Licklider envisioned no such exalted role for librarians. He hoped to build a more democratic system, one in which any number of nonexpert users could interact with computer terminals designed to access a shared "fund of knowledge." As the Defense Department continued to pour money into the ARPA, which in 1972 was renamed the Defense Advanced Research Projects Agency (DARPA), teams began to pursue something very much like Licklider's vision. Although Licklider himself left the agency in 1964, a succession of DARPA-funded researchers continued to pursue his ideas of a global network in an effort to build a more stable communications network. Popular Internet lore has it that the original Advanced Research Projects Agency Network (ARPANET) designers were trying to build a network capable of withstanding nuclear war, but in fact they simply wanted to build a better network (the nuclear war fable stems from the conflation of the ARPANET project with an unrelated RAND Corporation study during this period).[5] Researchers began stitching together a network of four major computing centers equipped with Interface Message Processors (IMPs), designed by a group of researchers working under contract at the technology consulting firm Bolt, Beranek and Newman.

In 1973, Vinton G. Cerf and Robert E. Kahn developed a pair of networking protocols known as TCP/IP (Transfer Control Protocol/Internet Protocol) that allowed for two or more computers to communicate with each other over the still-primitive ARPANET, by sending and receiving small bundles of data called "packets." Each packet included a numeric routing address—a kind of digital mailing label—that allowed it to move independently of the others, skipping around any points of congestion or technical failure to find the most efficient path through the network. This approach to networking marked a significant departure from previous attempts, which had largely relied on the traditional so-called client-server architecture—in which smaller, less-powerful terminals are tethered to a central server. In short, the Internet (a term first used by Cerf and his colleagues in a 1974 technical specification)[6] treated each machine as a peer. By establishing a networked system devoid of any single coordination point, designers ensured that the network would not only withstand global catastrophe, but also resist centralized control.

For most of the Internet's first two decades—the late 1960s to the late 1980s—access remained limited to the military and academic worlds that had spawned it, gradually extending to include more and more universities and other research institutions. Most of the traffic flowing across the network remained scientific and technical—scholarly papers, research data, and related discussions. In 1991, the National Science Foundation decided to allow commercial traffic—a turning point that would ultimately transform the Internet into something very much like the global network that Licklider had envisioned.

That same year, Tim Berners-Lee and his partner, Robert Cailliau (a notable Belgian information scientist), released the first public version of the World Wide Web, while working at the CERN particle physics accelerator laboratory in Switzerland. That system—the one many of us now use every day—consisted of a global network of wired and wireless devices that would allow anyone, anywhere in the world, to retrieve and display content on a screen capable of projecting text, images, and other audiovisual material.

Of course there are parallels to the Mundaneum, although none of those mentioned in this brief history of the Internet seem to have known of Paul Otlet. As remarkable as those similarities may seem, the two systems also differed in fundamental ways. Putting aside the obvious technological differences—Otlet worked well before the age of digital computers, after all, when the available state-of-the-art technologies included microfilm, index cards, and telegraph wires—the modern Web also reflects a fundamentally different architectural philosophy, at least in principle. "There is no 'top' to the World-Wide Web,"[7] declared the WWW Consortium in its early days—meaning no central server or authoritative body.[8] The Mundaneum, by contrast, would have relied on a central administrative organization coordinating the work of local, national, and international organizations. It would also have enforced a consistent set of formats and classification schemes, while employing a small army of trained "bibliologists" to collect and synthesize information from every possible source. The Web, by contrast, relies largely on search engines—and, increasingly, social media—to accomplish that task. In such an environment, the notion of "universal" classification seems like an enormous act of cultural hubris, if not outright imperialism. Such an uncontrolled, chaotic environment would likely have struck Otlet as the grandest kind of folly; yet some Web pundits celebrate the Web's fundamental disorder as its greatest strength.[9]

Futuristic though it was, and sometimes fabulously so, Otlet's Mundaneum was a product of its time, coming to existence in an era when the possibility of assessing and conveying "universal" truth still seemed achievable. While the Internet's pioneers may have aspired at the outset to accomplish a similar goal of unifying the world's knowledge, the Web has turned out to be something like a global free-for-all. While the Web's openness has fostered a great flowering of human expression, creativity, and opportunities for collaboration and social bonding, it has also left a vast body of human intellectual energy floating unmoored in the ether. Just as Otlet's Mundaneum was a product of its time, the Web is a product of ours. The Web's architecture—flat, open, and highly distributed—stems not only from its adherence to a particular set of technical protocols rooted in the U.S. military establishment, but from a particular strain of thought about the structure of hypertext systems that emerged in the Anglo-American computer world in the years after World War II.

Most contemporary accounts of Web history trace its conceptual origins to a 1945 essay by Vannevar Bush entitled "As We May Think." First published in the *Atlantic Monthly* and later in *Life* magazine, the essay describes a fictional machine called the Memex, consisting of a single desk with two screens and microfilm storage, providing access to a vast collection of stored documents.[10] Published fourteen years after Emanuel Goldberg first unveiled his Statistical Machine, the Memex drew on Bush's earlier experiments working with microfilm-reading machines in the 1930s.[11] Bush envisioned his new proposed device as far more than a microfilm reader, however. He saw it as capable of allowing users to search and retrieve information from a large body of stored information; create connections between them; add their own annotations; and ultimately create new, encyclopedic forms of knowledge that would emerge over time through their collective efforts.

Bush felt that scholars and other intellectual workers needed new tools to cope with the burgeoning problem of information overload, a problem he saw growing in tandem with the increasingly narrow focus of academic disciplines. "There is a growing mountain of research results," he wrote in "As We May Think." "But there is increased evidence that we are being bogged down today as specialization extends. The investigator is staggered by the findings and conclusions of thousands of other workers—conclusions which he cannot find time to grasp, much less to remember, as they appear." Bush, like Geddes, also felt that the compartmentalization of scholarship threatened to constrain humanity's intellectual potential, by forcing it into the same rigid, deterministic structures of academic departments. Traditional libraries had outlived their usefulness, failing to adapt to the increasingly unwieldy size of their collections, and growing too wedded to "the artificiality of systems of indexing." "The human mind does not work that way," he argues in the essay. "It operates by association. With one item in its grasp, it snaps instantly to the next that is suggested by the association of thoughts, in accordance with some intricate web of trails carried by the cells of the brain."

In order to support the organic processes of human thought, Bush believed that the world needed a fundamentally new technological platform for managing the outpouring of recorded data. "The summation of human experience is being expanded at a prodigious rate," he wrote, "and the means we use for threading through the consequent maze to the momentarily important item is the same as was used in the days of square-rigged ships."[12]

Bush had experimented with microfilm at some length, including a disappointing effort called the Rapid Selector, a malfunction-prone microfilm-based search-and-retrieval system. Unable to overcome the technical hurdles to building a better machine with the available technology at his disposal, Bush instead began to sketch out the contours of a new "concept car" machine, one designed not as a mechanical blueprint but rather as a conversation starter to encourage further exploration. In that respect, the Memex has proved a durable icon, long since attaining the status of platonic object in the computer science world.

Otlet shadows much of this, and there is considerable overlap in their thinking, not just in the potential of the network to move beyond the traditional library but also in the way it would require a new kind of intellectual work. It would not require computer programmers—the Memex was not programmable—but rather a new kind of writer-researcher performing a task not far removed from Otlet's "bibliologists." "There is a new profession of trailblazers," writes Bush, "those who find delight in the task of establishing useful trails through the enormous mass of the common record." "Thus science," he adds, "may implement the ways in which man produces, stores and consults the record of the race."[13]

MEMEX in the form of a desk would instantly bring files and material on any subject to the operator's fingertips. Slanting translucent viewing screens magnify supermicrofilm filed by code numbers. At left is a mechanism which automatically photographs longhand notes, pictures and letters, then files them in the desk for future reference.

Alfred D. Crimi's rendering of the Memex in *Life*, September 10, 1945. Courtesy of Joan D'Amico (Alfred and Mary Crimi Estate).

The Memex itself appears, at first glance, not far removed from Otlet's Mondotheque—and the two were conceived of only four years apart. Both machines relied on microfilm for storing large collections of texts. Both were to allow users to collect, collate, and annotate material, then share it with others. And both called for speech to be transformed into text. But they differed in several important respects. Otlet had imagined a system that encompassed all kinds of different media types—including film, photographs, and radio—whereas Bush saw his primarily as a machine for recording written ideas, as well as for more prosaic data storage functions, such as keeping track of accounts in a department store. And, as we've seen, Bush also eschewed the kind of top-down hierarchical classification system that Otlet had championed with the Universal Decimal Classification, which would have enabled the Mondotheque to retrieve precisely targeted information from a repository of data stored on index cards. Bush harbored lingering suspicions about such management strategies, with their rigid library science classification systems and—perhaps more important—bureaucratic overhead.

The Mondotheque also differed from the Memex in one critical respect: networking. For all the historical attention that has been accorded the Memex as the conceptual forerunner of the World Wide Web, the machine Bush envisioned was, in effect, a standalone terminal; it had no built-in facility for communicating with other machines. Otlet designed the Mondotheque, by contrast, to serve as a node in a globally distributed, wired and wireless network.

Bush's article, published in the giddy aftermath of World War II, captured the imagination of the American public. The article circulated widely and entered the popular consciousness just at the time that the earliest digital computers were starting to take shape. Those efforts, such as the ENIAC and MARK II computer, marked the beginning of the Cold War–era military-industrial research that would focus the intellectual firepower of the young computing industry primarily on scientific and military research projects. These involved problems of brute-force calculation and data storage rather than the more humanistic conundrums of knowledge management that Bush, Otlet, Wells, and others had pondered. But Bush's ideas percolated in the background for years, until a new generation of information scientists would build on his vision with the first computer-based hypertext experiments of the 1960s. Years later, Licklider would dedicate his report, *Libraries of the Future*, to Vannevar Bush.

On December 9, 1968, Douglas Engelbart took the stage before a packed house at Brooks Hall Auditorium in San Francisco to give his first public demonstration of the oN-Line System (NLS). Dressed in the white shirt of a working engineer, the soft-spoken former Navy telegraph operator demonstrated a working model of a system that struck many of the idealistic San Francisco counterculture types in attendance as representing nothing short of a revolution in human consciousness.

Equipped with a video monitor, keyboard, and central processor, Engelbart's demo included applications for word processing, sending messages between users, and even building links from one document to another. Stewart Brand (of the *Electric Kool-Aid Acid Test* and *Whole Earth Catalog* fame) manned a video camera trained on Engelbart's on-stage keyboard, while Engelbart proceeded to show a working prototype of a fully functional hypertext system, including a word processor, video and graphics displays, and the ability to link one document to another, all connected to another computer in Menlo Park by a 1,200-baud modem. The system also featured a never-before-seen device for pointing at objects on the screen: a small wooden box with wheels attached to the bottom that Engelbart eventually dubbed the "mouse."

Engelbart had encountered Bush's essay while stationed in the Philippines after World War II. The article made a lasting impression on him. In the years that followed, he pursued training in computer science, then played a role in a project at the Stanford Research Institute, under the tutelage of Licklider. The project involved what he called "augmenting human intellect." Although his project was born of Department of Defense–funded research at SRI, Engelbart evinced little interest in the mathematical and statistical problems that were receiving so much attention in the computer science world. Although an electrical engineer by training, he also had a strong humanistic streak, and he felt that computers should help all kinds of working professionals do their jobs—diplomats, executives, social scientists, lawyers, and designers, as well as scientists, like biologists and physicists—who at the time had little contact with computers, which still remained by and large relegated to the back office. The vision he ultimately pursued would look a lot like the modern personal computer, accessible to all.

"I had grandiose ideas about computers and screens," Engelbart later said, freely admitting that in the view of some of his professors he probably seemed a little "out there." In 1961, a colleague at SRI put him in touch with the International Foundation for Advanced Study, which had started to administer some of the earliest LSD experiments; the researchers were specifically interested in assessing the drug's effect on scientists and engineers, to explore whether the drug helped them achieve conceptual breakthroughs with difficult problems. During his first session, the former Navy officer sat in almost complete silence, staring at a wall for the length of the entire experiment. But he found the experience intriguing and suggested to the researchers that they try another experiment—this time with his entire team. "If you really believe we can be more creative, why don't we try this as a group and see if we can actually invent something?"[14] The results of the next session seemed underwhelming; Engelbart came up with an idea for a "tinkle toy" for helping to potty train young boys. Over the next few years, however, Engelbart's team began to expand their horizons, working amid the San Francisco Bay area's petri dish of cultural experimentation.

When Engelbart took the stage in San Francisco that day in 1968, many of the attendees felt they had experienced a revelation. As *New York Times* reporter John Markoff puts it, "Every significant aspect of today's computing world was revealed in a magnificent hour and a half."[15] Some members of that audience became enthusiastic converts to the digital revolution. Brown University computer science professor and early hypertext pioneer Andy van Dam was there and subsequently dubbed the event "the Mother of all Demos." Also in attendance were a few key members of the original NLS team, who migrated over to Xerox's PARC research division under the direction of Alan Kay, with whom they began developing the first true personal computer, the Alto. Stewart Brand, who was of course there, later brought the novelist and ur–Merry Prankster Ken Kesey over to look at the system; Kesey promptly dubbed it "the next thing after acid."[16] By the early 1970s, a "People's Computer Center" had appeared in Menlo Park, providing access to rudimentary computer tools that would allow customers to play games or learn to program. In the mid-1970s a young Steve Jobs (another LSD experimenter) first caught a glimpse of the graphical user interface (GUI) at Xerox PARC, soon licensing the software that would shape the subsequent trajectory of the Macintosh operating system and influence the design of the personal computer operating systems that most of us still use.

The counterculture of the 1960s and 1970s would play a formative role in shaping the personal computer revolution that followed. A belief in personal liberation, political empowerment, and world peace shaped the attitudes of the early personal computer hobbyists. Those values also found strong purchase in the academic computing centers, where the Internet first took root. In contrast to the top-down, centrally managed world of mainframe computers that still dominated in the corporate, military, and government power centers that funded the bulk of computer purchases, the new breed of counterculture programmers valued free expression and self-determination. "Half or more of computer science is heads" (meaning, roughly, hippies), wrote Stewart Brand in a landmark profile of the Bay Area computer science scene for *Rolling Stone* magazine.[17] Imbued with an ethos of individual freedom and self-expression, many of the early acolytes of the digital revolution—like Brand, Kevin Kelly, Howard Rheingold, and others—came of age during this period when top-down schemes were seen as tools of suppression and control, administered by "the Man."

Those counterculture idealists all opposed war and believed in the possibility of emerging technologies to usher in a new age of planetary consciousness and spiritual enlightenment. They diverged, however, in the paths they chose to pursue those exalted states. The anti-institutional animus of the 1960s counterculture stands in stark contrast to the enthusiasm felt by Paul Otlet and others for international associations, governing bodies, and associated rules and procedures. They believed in personal liberation and the disruption of what they considered anachronistic, calcified power structures. Otlet and others of the prewar generation, by contrast, placed great faith in certain kinds of organizational structures—specifically, international associations and pan-world government—that Otlet, for one, saw not as tools of bureaucracy and oppression but rather as the engines of great cultural and social reform, and ultimately as catalysts for humanity to achieve its true potential. Otlet never embraced the American ideal of individualism and personal liberation. His particular style of utopianism drew him more toward the universal than the individualistic.

Otlet's closest intellectual cousin might be Ted Nelson, in whom the idealistic ethos of the 1960s computer counterculture found its fullest expression. A gifted and iconoclastic thinker, Nelson built on the ideas of Bush and Engelbart and proposed an even more individualistic, humanistic vision of networked computing—one that would directly inspire Tim Berners-Lee's ideas for the World Wide Web. A former Harvard sociology student and onetime filmmaker who took an interest in computers, Nelson—like Otlet and Geddes—believed in a vision of technology rooted deeply in the social sciences. He once even dubbed himself a "systems humanist."[18] Taking his inspiration from the architecture of Frank Lloyd Wright and Otlet's collaborator Le Corbusier, as well as Bertrand Russell and Alfred North Whitehead's *Principia Mathematica* (a book that attempted to synthesize the entirety of known mathematics into a small number of basic concepts), Nelson began to pursue a postmodern vision of what computers might become.

"The public had been told that computers were mathematical, that they were engineering tools," Nelson recounts in his autobiography, *Possiplex*. "This misstated things completely. The computer was an all-purpose machine and could be whatever it was programmed to be. It had no nature; *it could only masquerade*" (emphasis Nelson's). Nelson also felt that the interfaces of contemporary computers reflected a too-unquestioning acceptance of authoritative organizational structures that, in his view, ultimately held computers back from their liberating potential. "The implicit choices made all over the paper world—by librarians, office supervisors, clerks, everybody—had to be made explicit and locked into software."[19] Nelson began to envision what he considered a better way, one geared more toward individual self-expression rather than fulfilling the process imperatives of hierarchical organizations.

Fascinated by the performative aspect of computers, Nelson began developing a vision of a computing environment for non-technical users. Using such a system, anyone could publish material without relying on institutional gatekeepers (like publishers). Perhaps more important, he imagined that such an environment would allow users to create entirely new forms of expression, building on a principle he has since called "profuse connection," which he defined as the "whole problem of abstraction, perception and thought." "Trying to communicate ideas requires selection from this vast, ever-expanding net," he wrote. "Writing on paper is a hopeless reduction, as it means throwing out most of the connections."[20] Animated by this fundamental insight about the importance of expressing connections between documents—a concept not far removed from Otlet's monographic principle—Nelson started to develop a system in which a large body of textual information might be collected, organized, and repurposed into new forms, allowing users to create visible links between documents. In 1965, he wrote a paper in which he coined the term "hypertext."

Otlet had used the term "Hyper-Documentation" more than three decades earlier, but he was referring to mixed-media, multi-sensory experiences. Nelson meant something different: a new kind of deeply interlinked electronic document. Still, van den Heuvel suggests that Otlet's approach to classification as a documentary language anticipates Nelson's concept of hypertext, insofar as it encompasses nonsequential and multidimensional relationships[21] or, as Nelson put it, "text that branches and allows choices."[22]

Nelson maintains that he spent his first five years thinking about interactive text systems in conceptual isolation, with no idea that others had developed similar concepts. Nelson's earliest proto-hypertext experiments bore a striking resemblance to Conrad Gessner's: using card files, notebooks, scissors, and paste. Over time he hoped to create "the *dream* file" (emphasis Nelson's), a sophisticated writing and filing system designed for an author, "holding everything he wanted in just the complicated way he wanted it held, and handling notes and manuscripts in as subtle and complex ways as he wanted them handled."[23]

In addition to thinking deeply about the problems of managing large collections of text, Nelson also envisioned how other forms of expression might take shape over time: "hypergrams," "hypermaps," and "hyperfilms" or "branching movies." Over the coming decades, Nelson self-published a series of books with titles like *Computer Lib* and *Dream Machines,* in which he articulated his evolving ideas about networked computing. Relying on metaphors, kinetic illustrations, and inventive neologisms ("Thinkertoys," "Indexing vortexes," "Window sandwiches"), Nelson's books attracted a cultlike following, among them a handful of influential early programmers who were drawn to this radical and highly personalized form of networked computing.

In *Dream Machines,* published in 1981, Nelson first unveiled his plans for a system he called Xanadu, named after Samuel Taylor Coleridge's opium-fueled depiction of Kubla Khan's "pleasure dome." He described the system as "a world-wide network, intended to generate hundreds of millions of users simultaneously for the corpus of the world's stored writings, graphics and data." That system would evolve into a "universal data structure to which all other data structures will be mapped," enabling large numbers of users to collaborate in creating and sharing their intellectual output in text, graphic, and other digital forms, with "promiscuous linkage and windowing among all materials."[24]

There are sturdy strands of connection between Otlet, Bush, and Nelson. While Nelson may have developed his earliest concepts by himself, by the time he published his landmark paper, he recognized a debt to Bush, quoting at length from "As We May Think" in his landmark 1965 paper, discussing his concept of associative trails and two-way links between documents.[25] While acknowledging the debt to Bush, however, he also pointed out that Bush's Rapid Selector (a real-world microfilm reader that preceded the fictional Memex) "is not suited to idiosyncratic personal uses."[26]

"Idiosyncratic" scarcely begins to capture the dizzying, strident tone of Nelson's writing. At his best, Nelson bristles with sparkling and original insights about the possibilities of electronic documents; at his worst, he comes across as petulant and vindictive, blaming his failures on others and ranting against institutional authorities. In this respect Otlet and Nelson seem like kindred spirits indeed. Nelson's writing shares much of the same utopian zeal that characterized Otlet's work, but as noted he did not share Otlet's faith in institutional authority. Otlet conceived of his systems at a time when the center was not holding, when a new centralized authority seemed to promise peace and stability over the warring nation-states. Nelson, by contrast, wrote during the rise of the counterculture, when the Cold War ascendancy of the U.S. military-industrial complex threatened to subsume the promise of computers for personal liberation. Nelson looked suspiciously on any attempts at imposing centralized hierarchical systems. Indeed, he regularly railed against hierarchies of all kinds. "From earliest youth I was suspicious of categories and hierarchies."[27] That distrust extended from file systems to bureaucratic structures of any kind. He heaped particular scorn on large institutions, especially universities, which in his view perpetuated a warped hierarchical understanding of the world of human knowledge. The modern university, he wrote in his book *Literary Machines,* "imbues in everyone the attitude that the world is divided into 'subjects'; that these subjects are well-defined and well-understood; and that there are 'basics,' that is, a hierarchy of understandings."[28]

While Otlet's Mundaneum might seem on the surface to resemble exactly the kind of top-down entity against which Nelson so frequently railed, upon closer inspection it would have incorporated many of the same principles of conceptual flexibility and multidimensionality that Nelson embraced. The Universal Decimal Classification is far from a simple top-down conceptual hierarchy; it is explicitly nonlinear and multidimensional.[29] It is too easy to conflate the classification scheme with the centrally managed institutional organization that Otlet envisioned to support it. Nonetheless, what Nelson ultimately envisioned would likely have appalled Otlet in its anarchic nature; and Otlet's new world government would almost certainly have repulsed Nelson in its reliance on a central office to enforce documentary standards.

For all of his railing against top-down hierarchies and systems of control, however, Nelson shared Otlet's concern for preserving the integrity of source materials. And for all his countercultural distrust of authority, he understood the importance of intellectual property controls, proposing a structure for clearing copyrights and facilitating payment that would make it easier for content creators to repurpose material without having to worry about securing permissions in advance (the Creative Commons licensing standard is an attempt to address this issue to some extent). To ensure the provenance of documents, he called for stabilized content addresses that would prevent rampant duplication of material. In contrast to the Web, which makes cutting and pasting as easy as pointing and clicking, in

Nelson's environment each document would exist in only one location in the system. For example, Lincoln's "Fourscore and seven years ago" would, in Nelson's system, have a consistent addressable identifier: ["Gettysburg Address" ID] 0–25.[30] This notion of classifying the contents of documents by means of a deep indexing tool seems philosophically compatible with Otlet's Universal Decimal Classification.

Otlet had, as we've seen, imagined universal books collecting the available material on a given subject in all available media—what Bush called "new forms of encyclopedias"—and Nelson was clearly working along the same lines. He proposed new kinds of higher-order distillations of original material taking shape as "hyperbooks" and "grand systems," consisting of "'everything' written about the subject, or vaguely relevant to it, tied together by editors." These editors would have much in common with Otlet's bibliologists.

Though ultimately never launched, Nelson's Xanadu system exerted a powerful influence on the subsequent trajectory of the Web, whose present incarnation continues to appall him. In recent years, he has decried it as "a nightmare honkytonk prison, noisy and colorful and wholly misbegotten." By choice, Nelson has remained an outsider to the industry where he is seen in some quarters as an eccentric crank, in others as a crazy genius and provocative teller of truths. Genius or crank, he has emerged as an influential critic of modern computing, and his unapologetic rants may yet serve as a useful corrective to the self-serving triumphalism of corporate titans like Apple, Microsoft, and Google.

12

Entering the Stream

Although researchers had been experimenting with hypertext and networked information systems for decades—from Otlet's Mundaneum to Bush's Memex, Engelbart's oN-Line System, Nelson's Xanadu, and so on—it was not until the 1990s that the dream of a global knowledge network finally seemed within reach. When the National Science Foundation lifted the final restrictions on commercial use of the Internet in 1991, the decision came just as a promising new hypertext program emerged from the CERN particle physics research center in Switzerland.

CERN researcher Tim Berners-Lee had been working for several years on a system that would allow the center's researchers to share information with each other more easily. Berners-Lee has freely admitted that his ideas were influenced by Ted Nelson; he had also immersed himself in the increasingly active hypertext research community of the late 1980s. In his 1989 proposal to his manager at CERN, Mike Sendall, Berners-Lee grounded his argument in the challenge of knowledge management in a large organization producing reams of scientific data. "If a CERN experiment were a static once-only development, all the information could be written in a big book," he wrote. But static books would never meet the needs of this large and constantly evolving organization. "Keeping a book up to date becomes impractical, and the structure of the book needs to

Vague but exciting ...

CERN DD/OC

Information Management: A Proposal

Tim Berners-Lee, CERN/DD

March 1989

Information Management: A Proposal

Abstract

This proposal concerns the management of general information about accelerators and experiments at CERN. It discusses the problems of loss of information about complex evolving systems and derives a solution based on a distributed hypertext sytstem.

Keywords: Hypertext, Computer conferencing, Document retrieval, Information management, Project control

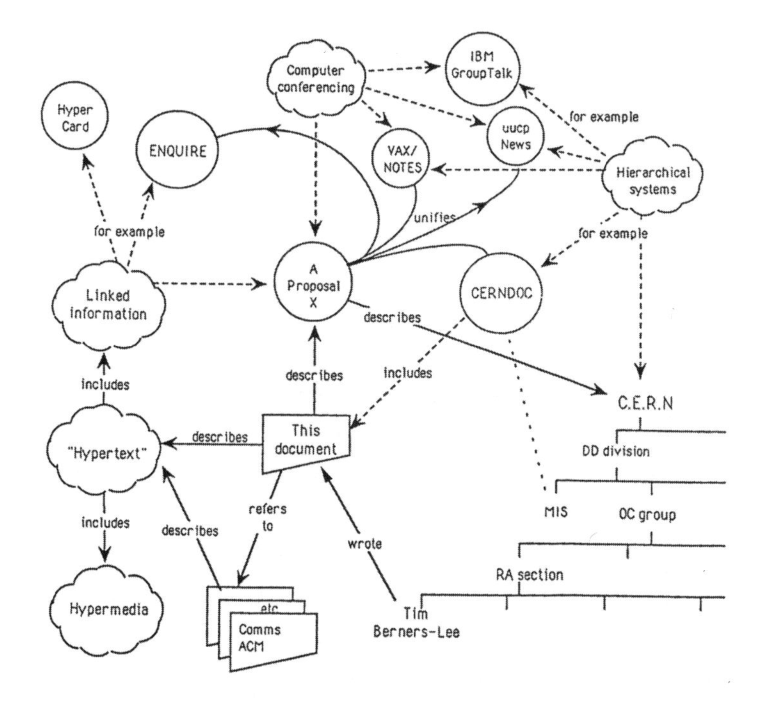

Tim Berners-Lee's original proposal for the World Wide Web, 1989. ©CERN.

be constantly revised." The solution to the problem of a book's limits was the network of communication that already existed within CERN. More important, he foresaw that the challenges he had identified there might have greater implications. "CERN is a model in miniature of the rest of the world in a few years' time."[1]

"Vague, but exciting," wrote Sendall, in response to the proposal. That cautious note of encouragement gave Berners-Lee all the approval he needed to move forward with creating the first version of the World Wide Web. In the years that followed, Berners-Lee's instincts proved exactly right. At first Berners-Lee called the system a "Mesh"; it would be another year before he stumbled on the name that finally stuck: World Wide Web. The Web has long since swept aside or co-opted any competing attempts at building global hypertext systems. Nonetheless, it did not come out of nowhere. Berners-Lee worked within the context of a hypertext movement that had grown up in the computer science world of the 1980s. In his initial paper he mentioned the pioneering work of Ted Nelson, as well as the many contributions of researchers involved in two hypertext conferences in 1987 and 1988. Another hypertext pioneer, Andy van Dam, had been conducting important hypertext experiments at Brown University starting in the 1960s. Several European teams were also pursuing alternative visions of hypertext, notably Wendy Hall's Microcosm project at the University of Southampton in England, and Keith Andrews's Hyper-G project at Austria's Graz University of Technology.

Berners-Lee acknowledged that the Web lacked some of the sophisticated features of other systems. "Much of the academic research is into the human interface side of browsing through a complex information space," he wrote. "Whilst the results of the research are interesting, many users at CERN will be accessing the system using primitive terminals, and so advanced window styles are not so important for us now."[2] The document-sharing tool he had envisioned for his colleagues quickly showed its limits. The first release supported only text; support for images would come later, as would other media formats like audio and video. Perhaps more important, the Web provided a simple platform for linking one document to another—but not for controlling a single version of that document, managing the identity of the user, or ensuring copyright. Those limitations might have carried limited consequences in the academic environment of CERN, but they would hamstring the Web for years to come when it eventually found its way into the commercial marketplace.

The Web reflects its technological and cultural pedigree, relying on a set of standards that has now achieved canonical status in the computer science world: namely, the HyperText Transfer Protocol (also known as HTTP, as in the ubiquitous "http://" prefix that appears in most browser windows) and the Uniform Resource Locator (URL), a standard for managing the addresses of individual machines. Both standards are "open," meaning they do not belong to any particular company or government entity; rather, these standards evolve over time under the guidance of international standards committees coordinated by the World Wide Web Consortium (W3C). The reliance on an international body to ensure the open flow of information carries faint echoes of Otlet's hopes for a worldwide administrative organization. But while the present-day W3C sets technical standards on behalf of billions of Web users, it stops far short of serving as an institutional arbiter of facts—let alone fulfilling the Otletian dream of uniting the world's governments and religions. In practice, the W3C's mission hews more closely to the Nelsonian principle of decentralization. The simplicity and openness of Web standards have lowered the barriers to entry for untold millions, allowing more people to start businesses, making it cheap and easy to distribute recorded material, and allowing for new forms of social organization to spring up in the form of self-organizing social networks. Yet the Web's simplicity has a downside: the lack of stable identity management, intellectual property controls, or reliable archiving capabilities. Coupled with ongoing concerns about worms, viruses, and other forms of "malware," the Web thus seems to exist in a state of perpetual anarchy.

In 1998, just as the World Wide Web was establishing itself as a transformative force in the world economy, Tim Berners-Lee wrote an essay reflecting on his conversion to the Unitarian Universalist church. The church has its roots in the universalist movement of the late nineteenth century—the same spirit of pan-global utopianism that had animated Otlet, La Fontaine, Le Corbusier, and so many others. Berners-Lee wanted to stress the philosophical parallels between his original vision for the Web and the guiding principles of Unitarian Universalists. Indeed, he allowed that the same spirit that led him to create the Web had ultimately attracted him to the church. While he cautioned against overstating the comparisons—"Let's take this all with a pinch of salt"—he nonetheless outlined some intriguing conceptual overlaps. "So where design of the Internet and the Web is a search for set of rules [*sic*] which will allow computers to work together in harmony, so our spiritual and social quest is for a set of rules which allow people to work together in harmony," he wrote in his 1998 essay "The World Wide Web and the 'Web of Life.'" The Web, then, reflects a set of idealistic assumptions about social and spiritual relationships that might yield tangible improvements in human social, political, and economic relationships. Like the Unitarians, he wrote, the Internet community valued decentralization, tolerance, independence, truth, and hope.[3] "The essential property of the World Wide Web is its universality," he wrote. Otlet saw universalism as a political and spiritual ideal, a general principle.

Berners-Lee was taking a more literal-minded view of the term. He interpreted "universality" to mean as many reflections of human thought as possible—"one light, many windows," as the influential Universalist minister and theologian Forrest Church once put it in a sermon[4]—rather than trying to define a single authoritative statement of universal truth. "Web technology...must not discriminate between the scribbled draft and the polished performance," he wrote, "between commercial and academic information, or among cultures, languages, media and so on. Information varies along many axes." For Berners-Lee, the Web was to be an inclusive, open, and democratic arena rather than the centrally managed environment that Otlet imagined. Despite the differences between them—centralization versus decentralization—Berners-Lee was building on Otlet's democratic instincts. The two men also shared a spirit of optimism about the possibilities of networks as agents of positive social change. Otlet hoped to encompass the entirety of human knowledge in the Mundaneum. Berners-Lee believed the test of the Web would be its ability to encompass the broadest possible spectrum of human thought. "It must be able to represent any thought, any datum, any idea, that one might have," he wrote in that same essay. "So in this way the Web and the [Unitarian Universalist] concept of faith are similar in that both serve as a place for thought, and the importance of the quest for truth."[5]

That optimism notwithstanding, Berners-Lee himself has been quick to acknowledge some of the Web's fundamental limitations. As early as 1996 he appeared before the W3C and lamented the lack of two-way authoring in modern Web browsers.[6] In 2000, he published a paper in *Scientific American* in which he laid out his vision for a more orderly version of the Web that would allow computers to exchange information with each other more easily. He dubbed the project the Semantic Web, an umbrella term for a collection of technologies aimed at making the Internet more useful by imposing more consistent structures on data that can then be exchanged automatically between machines. He envisioned a "Web of data" designed primarily to foster the automatic exchange of information between computers, to allow any number of applications to search, retrieve, and synthesize data drawn from disparate sources. "The Semantic Web will bring structure to the meaningful content of Web pages," he wrote, "creating an environment where software agents roaming from page to page can readily carry out sophisticated tasks for users."[7] Substitute "bibliologists" for "software agents," and that same description could just as easily apply to the Mundaneum. Otlet also considered the possibility of engineering mechanical means of indexing and reassembling data from multiple sources, as in his microfilm experiments with Robert Goldschmidt. The environment that Otlet envisioned bears other similarities to the Semantic Web. By predicating its future on ontologies, handcrafted maps of topical relationships, the initiative shares the same spirit of expert knowledge systems that characterized Otlet's work with the Universal Decimal Classification. Van den Heuvel, for one, argues that Otlet's framework not only resembles the hyperlinked structure of the World Wide Web, but also presages some of the more advanced linking strategies of the Semantic Web.[8] Otlet's Monographic Principle provided a framework for breaking down documents and other forms of media into component parts, then recombining them into new formats using the multiple link dimensions afforded by the Universal Decimal Classification. Both systems allowed for the collection and classification of documents, linking them programmatically by exchanging metadata to create pathways of association from one piece of information to another. However, van den Heuvel also points out that while both systems share more than a few surface-level similarities, Otlet's network differs importantly from the Semantic Web and the WWW in terms of their goals and expectations.[9]

Berners-Lee's Semantic Web relies on a data model known as the Resource Description Framework (RDF) to encode these relationships; RDF hinges on the use of so-called triples to denote subject-predicate-object relationships. For example: The Beatles (subject); member (predicate); John Lennon (object). These triples can then be stitched together to create relationships with other bits of data. The triple could then easily be associated with another triple sharing a common element: Beatles (subject); album (predicate); *Abbey Road* (object).

As van den Heuvel argues, Otlet's Universal Decimal Classification serves much the same purpose as RDF triples. By expressing the relationships between classification numbers, Otlet's system establishes semantic links that can connect one concept to another; then, using the auxiliary tables of the Universal Decimal Classification (such as "+," "/," and ":"), the system can describe the nature of that relationship between topics, languages, geographical locations, and so forth.[10] Using this framework, one could imagine an application that would allow a user to explore an entire data universe of

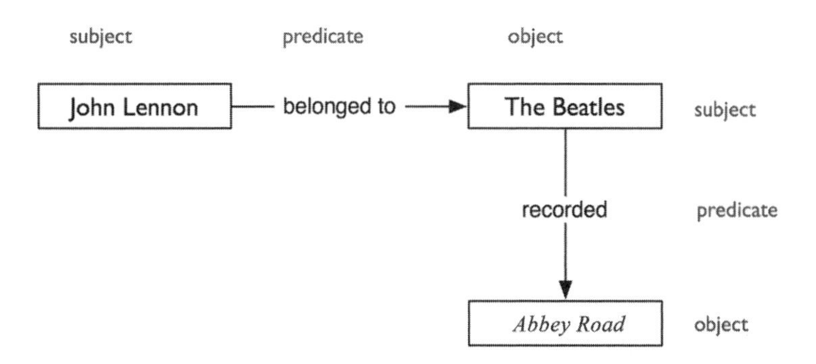

RDF triples. Illustration by Alex Wright.

information about music and musicians by following trails of association from one data point to another. Other applications could then merge that data with other sources (a biographical database, a catalog of recordings, and a fan site, for example) to create new synthetic displays that gather content from multiple discrete collections. Otlet imagined that his bibliologists would synthesize information from multiple sources to give readers an overview of any given topic with links to support further exploration.

Critics have argued that Berners-Lee's Semantic Web amounts to an attempt at centralizing control in the hands of a few (after all, how many people could possibly claim the authority to determine the structure of its underlying ontologies?); they have also argued that it calls for a priesthood of computer programmers to maintain it. The Web, by contrast, has traditionally lent itself to amateur exploration—indeed, this is precisely why it has achieved such widespread consumer adoption. The question of who might "own" the Semantic Web has proved one of its thornier problems and points to an absence of systemic thinking about the attendant policy and organizational issues surrounding it. Technology alone, it seems, is not enough. Without an administrative operation equal to the task of organizing the world's knowledge, such an endeavor seems doomed to incompletion.

In a 2004 essay for *The Edge,* an online salon founded in 1996 by author and literary agent John Brockman, supercomputer pioneer Danny Hillis described his own vision for a new learning environment rooted in Semantic Web principles, to create a so-called Knowledge Web in which "humanity's accumulated store of information will become more accessible, more manageable, and more useful." The environment would include carefully classified metadata on a wide range of topics, creating a rich universe of links uniting a large corpus of information. "Anyone who wants to learn

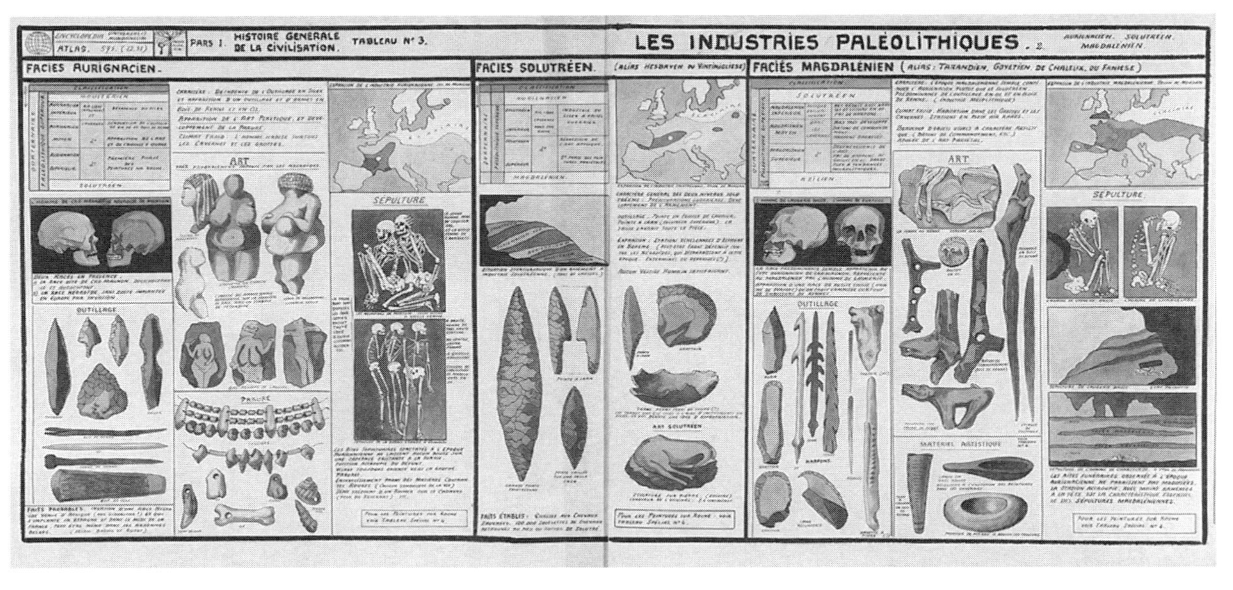

Paleolithic exhibit diagram, 1920s. Reproduced with permission from the Mundaneum.

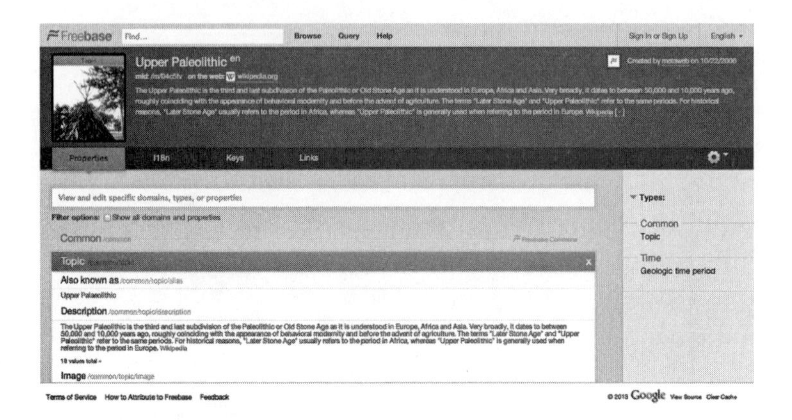

The same topic treated in Freebase, the open Semantic Web–based information repository. Licensed under Creative Commons 2.0.

will be able to find the best and the most meaningful explanations of what they want to know. Anyone with something to teach will have a way to reach those who what to learn [*sic*]," he wrote in the essay. "The knowledge web will make us all smarter."[11] Hillis's optimism for the possibilities of the Semantic Web eventually paid off. One year after writing that essay, he established a company called MetaWeb that created Freebase, which he characterized as an "open, shared database of the world's knowledge." In 2010, he sold the company to Google, where its structured snippets now often complement traditional keyword-based search results.

In recent years, the Linked Data movement has to some extent subsumed the Semantic Web initiative. Linked Data proposes more of a middle ground, in which ontologies might be derived programmatically from analyzing large data sets, rather than manually created by teams of experts.[12] This middle way approach might incorporate some of Otlet's ideas: a topical structure further refined by automated discovery, bidirectional linking, and the ability to extract content from static documents, then synthesize and interpolate it in new ways.[13]

In a widely circulated 2005 essay, "Ontology Is Overrated," Clay Shirky argues that projects like the Semantic Web were doomed to failure in the Internet age. "The more you push in the direction of scale, spread, fluidity, flexibility," he wrote, "the harder it becomes to handle the expense of starting a cataloguing system and the hassle of maintaining it, to say nothing of the amount of force you have to get to exert over users to get them to drop their own worldview in favor of yours."[14] Shirky suggests that in the open environment of the Web, command-and-control cataloging structures are doomed to failure. Instead, Web site owners are better served by resorting to user-driven classification systems, suggesting that "the only group that can categorize everything is everybody." In the end, in Shirky's view, it comes down to a "question of philosophy": "Does the world make sense or do we make sense of the world?" If you believe the former then any program runs the risk of being wrong. And if you believe the latter any system you devise will be artificial and dangerously top-down.

The essay distills a prevalent mindset on the Web: that cataloging and classification are antiquated models, that the truth is relative, and that the hive mind of the collective intellect will sort everything out. It's an appealing notion, but in practical terms crowdsourced classification, which is what this represents, may prove no less a fantasy than Otlet's dream of a world government managing the world's intellectual output. For several years in the mid-2000s, Web cognoscenti held out high hope for so-called folksonomies (a neologism coined by Thomas Vanderwal), participatory tagging systems in which users classify material en masse using open-ended keywords, rather than relying on expert cataloging systems. Despite a burst of initial enthusiasm among Web designers and developers, that movement largely failed to bear fruit. But that has scarcely stopped many Web promoters from proclaiming the end of traditional hierarchical systems.

Technological predictions rarely come true exactly as planned. In 1949, *Popular Mechanics* magazine issued a breathless prediction that computers of the future might one day weigh less than 1.5 tons. In 1977, Digital Equipment Corporation chairman and founder Ken Olson said he saw no reason why anyone would ever want a computer in the home. In 2004, Microsoft founder Bill Gates predicted the death of spam within two years. It should scarcely come as a surprise that a generation of inventors and entrepreneurs working more than a half-century ago should have failed to anticipate the Web taking shape precisely in its present-day form. What is surprising is that they envisioned any such thing at all, given the technological constraints of the day. And while we cannot replicate the systems they envisioned—predicated as they are on antiquated technologies and painfully outdated cultural assumptions—they nonetheless do hold out a number of specific, tactical ideas that could potentially offer interesting ideas, or perhaps points of departure, for new developments. After less than a quarter-century, the Web remains a developing technology.

As sites and applications have grown more sophisticated over the past two decades, some of the Web's inherent weaknesses have come into sharper focus. On the one hand, the Web has fueled an outpouring of human expression and fundamentally transformed the way businesses and organizations work. On the other, it has ushered in a tumultuous period for many so-called content creators, by dint of its lack of intellectual property controls, archival function, or identity-management features. The absence of such mechanisms has created a state of perpetual ontological anarchy. Most users therefore turn to search engines like Google, which rely largely on keyword analysis rather than classification systems to gauge the contents of Web sites—an ostensibly more "natural" strategy that has in practice begotten a cottage industry of so-called search engine optimization (SEO) experts who collect fees to help customers game the results. Meanwhile, the great miasma that is social networking has fostered a kind of fluid, conversational mode of communication that mimics the ephemerality of oral culture rather than the gravity and aspirations toward the permanence of traditional literate culture. Meanwhile, Web sites come and go like shooting stars, with companies spending millions of dollars on flashy interactive projects only to discard them a few months later, often leaving no trace of their existence.

Those shortcomings have also created the opportunities for smaller walled gardens to thrive, for instance, Apple, which has created its own tightly controlled ecosystem of devices, applications, and media—all revolving around its own software and e-commerce systems—or Facebook, which exerts considerable control over the structure and presentation of content. Even Google, ostensibly devoted to preserving the open Web, exerts its own corporate judgment in determining which sites are considered acceptable (Wikipedia is in; child pornography is out). While there may be perfectly sound legal or moral reasons for these decisions, the fact is that untold billions of Web users must rely on the judgment of a corporate oligarchy whose algorithms remain as closely guarded as state secrets. It is not for the consumer to know why a particular result appears at the top of a results page; such are the Delphic mysteries of the Google searchbot. Otlet's system, though similarly conscribed, at least aspired to an idealistic vision of revealing truth without serving a corporate profit margin.

While corporations increasingly dictate the experience of most Web users, the Web itself remains fundamentally an open platform. The W3C manages the technical standards of the network, but no organization governs the enterprise as a whole. Some pundits have celebrated the Web's fundamental lack of order as its great strength.

The philosopher and Web pundit David Weinberger makes a particularly eloquent case for preserving that ontological chaos, arguing that human knowledge is inherently fluid and that attempts at orchestrating it are doomed to failure. Celebrating the "new principles of digital disorder," Weinberger argues that "we have to get rid of the idea that there's a best way of organizing the world." He further argues in favor of making as much information as possible freely available, without trying to exert quality controls. Instead, users should have open access to everything and choose whatever seems useful. "Filter on the way out, not on the way in," he writes, predicting the emergence of what he calls a "third order" of knowledge, one not constrained by the physical limitations of paper, nor encumbered by layers of institutional gatekeepers. Instead, he argues for a maximalist approach: collect everything and sort it out later, since the online world provides "an abundance of access to an abundance of resources."[15] Pointing to examples like Wikipedia and Flickr, Weinberger argues that the old order of knowledge—the Aristotelian notion that knowledge was shaped like a tree—has fallen by the wayside and that the free-form structure of today's Web provides a more accurate depiction of culture, one that "better represents the wild diversity of human interests and thought."[16]

In an expression of positivist-style thinking of which Otlet would doubtless have approved, Weinberger also describes the evolution of recorded knowledge as a series of progressive stages. In the first order, books are books; in the second order, metadata enters the picture, so that readers can find the books they want by way of descriptive cataloging information; in the third order—the era in which we now live—data becomes metadata, as everything becomes searchable, and top-down classification systems are no longer necessary.

Just as Otlet argued for liberating information from the confines of books, then adding it to a central, searchable index, so Weinberger suggests a merging of books and documents transcending the limitations of the traditional book-bound orientation to the structure of human knowledge. Physical libraries can only put each book in a single location. The Web makes it possible to "put each leaf on as many branches as possible."[17] In other words: To create a classification with multiple entry points and cross-references—much the same thing that Otlet hoped to accomplish with the multidimensional classification techniques of the UDC.

Many Web pundits also point to the near-total lack of hierarchy underlying one of the Web's largest and most successful sites: Wikipedia. Now one of the most heavily trafficked sites on the Web, its success seems to vindicate the hopes of many early information scientists—going back at least as far as Conrad Gessner and stretching on through the work of Wells, Otlet, and others—who dreamed of a universal encyclopedia. Wikipedia appears on the surface to fulfill that ideal, by consciously emulating the experience of a traditional encyclopedia over the Internet. Moreover, unlike Google, Amazon, Facebook, and the other major sites, Wikipedia remains a nonprofit endeavor—again, a seeming vindication of Otlet's vision of the universal encyclopedia as serving a higher intellectual purpose than mere commercial enterprise.

For all its scope and reach, however, Wikipedia differs in critical ways from the networked encyclopedia that Otlet, Wells, and others imagined. First and foremost, it embodies those core Internet values of openness and transparency. Almost anyone can contribute a document or edit the work of others, tracking all changes and discussions in public, under the minimal supervision of volunteer administrators drawn from the user community, supplemented by a small core paid staff. While the site has, over time, implemented a basic level of administrative hierarchy governing its editing process (which remains carried out overwhelmingly by volunteers), it remains devoted to the principle that anyone can contribute, rather than merely a group of authorized experts and institutional gatekeepers.

Critics have argued that Wikipedia's lack of hierarchy has created an unstable and unreliable reference source, one that despite its impressive scope—with more than 29 million pages and 18 million registered users—cannot effectively vouchsafe the quality of its content at any given time. In hopes of addressing the lack of domain expertise and control, Wikipedia co-founder Larry Sanger briefly started a new competing encyclopedia known as Citizendium, in which expert curators would play a more dominant role. Launched with much public relations fanfare in October 2006, Sanger vowed "to unseat Wikipedia as the go-to destination for general information online." He hoped to do so by releasing a more "credible" online encyclopedia, one in which contributors would agree to use their real names and work under "gentle expert oversight" to create articles approved by peer review of topic experts. Alas, such Otletian devotion to truth and high editorial standards failed to carry the project past the cultural juggernaut of Wikipedia. The Citizendium project proceeded to fall far short of its lofty goals. By October 2011, the site had dwindled to fewer than 100 active members and 16,270 articles, of which only 164 had received approval from the editorial committees.[18]

For all of Wikipedia's apparent invincibility, the site suffers from several fundamental shortcomings, especially when compared to some of the early large-scale hypertext experiments that preceded it. Like much of the rest of the Web, it hews to an explicitly paper-based metaphor, in which authors edit pages much as they might do with a traditional bound encyclopedia. While this embrace of analog metaphors makes the site familiar and accessible to many readers, it also fails to "chunk" its contents into structured data that could support more dynamic interactions. Instead, it stores all information in flat, freeform text files with limited standardization. And while it does utilize a small army of volunteer editors (and a small paid staff), Wikipedia by and large embraces the open-source ethos of the Internet, in which anyone (in principle) can create or edit pages. Wikipedia does not employ any particular hierarchical classification of subjects, relying instead on the insertion of hyperlinks resembling the "associative trails" that Bush once envisioned. Thus, the site relies not on a top-down orientation to human knowledge, but on the bottom-up distribution of traffic from across the Web, largely via Google and other search engines.

Wikipedia has succeeded by insinuating itself into the flat, open architecture of the Web as a whole. But that openness comes at a price: it has no underlying classification system to guide users to related topics; instead they must rely on the good-faith efforts of contributors of varying levels of expertise and authority—a far cry from Otlet's notion of trained "bibliologists" working within a tightly controlled classification scheme. Those differences aside, Wikipedia nonetheless seems to suggest the durability of the encyclopedia as metaphor, and the enduring appeal of attempts at creating a "universal" knowledge system on a global scale.

What would Otlet have made of Wikipedia? He almost certainly would have approved of its encyclopedic ambitions and global reach. However, he probably would have looked skeptically at the notion of an open authoring platform in which anyone can post, edit, or delete material with minimal oversight. Perhaps he also would have seen a lost opportunity in the "flatness" of the collection and lack of an underlying classification system that would have enabled its contributors to collect, synthesize, and recombine its content in new forms. Ultimately, he likely would have seen Wikipedia as a meager effort at creating a Mundaneum-like environment, with a similarly utopian ambition but none of the control systems that he felt would be critical to ensuring the long-term success of such an encyclopedic undertaking.

Meanwhile, back in Brussels, a group of researchers has been pursuing a similarly ambitious—though less well known—attempt to create a global knowledge network. Under the guidance of Francis Heylighen, a small team has been working for more than two decades to build a new World Brain. Since its public launch in 1990, the Principia Cybernetica project has been attempting to build "a complete and consistent system of philosophy" by creating new, networked forms of recorded knowledge. "This philosophical system will not be developed as a traditional document, but rather as a *conceptual network*," wrote two of the project team members in a 1989 introduction to the project. "Using this structure, multiple hierarchical orderings of the network will be maintained, allowing giving both readers and authors flexible access to the whole system."[19] They hope to create a grand synthesis of philosophical work, the Principia Cybernetica, by breaking the entire spectrum of philosophical thought into "nodes" in the network. These "nodes" might include a chapter from a book, a paragraph from an essay, or even more narrowly targeted units of information—a notion not far removed from Otlet's biblion.

In a 2012 essay entitled "Conceptions of a Global Brain," Heylighen wrote about the historical pedigree of the concept underlying the Principia Cybernetica project. He identified three major strands of thought: the organicists—including August Comte (who inspired Otlet and Geddes) and Herbert Spencer (who shaped Borges's views on the relationship between individuals and society)—tended to see society as a living system, with information coursing through it like a vital fluid. Encyclopedists—in whose number Heylighen included Otlet and Wells, as well as Bush, Nelson, and Engelbart—aspired to create an archive of all human knowledge, freely available to anyone in the world. Finally, emergentists—like Teilhard de Chardin and the physicist Peter Russell—embraced the spiritual possibilities of the global brain, aspiring to a higher state of collective consciousness. While it seems unfair to characterize Otlet as strictly an "encyclopedist," given that his work clearly encompasses both organicist and emergentist strains of thought, nonetheless it credits him as one of the global brain's conceptual forebears. For Heylighen, however, none of these approaches will suffice on its own. Instead, he calls for what he terms "evolutionary cybernetics," integrating Darwinian concepts of natural selection with the cybernetic concept of emergent levels of understanding. Applying this model, Heylighen believes, will finally transform the chaos of the Web "into an intelligent, adaptive, self-organizing system of shared knowledge."[20]

That notion of an emergent, evolving system has found plenty of adherents in the years since the Web first started to command a broad public audience. As *Wired* executive editor Kevin Kelly put it: "The revolution launched by Netscape's IPO was only marginally about hypertext and human knowledge. At its heart was a new kind of participation that has since developed into an emerging culture based on sharing." Kelly goes on, however, to evoke the same strain of technological optimism that marked the essays of the early twentieth-century visionaries, predicting that "the ways of participating unleashed by hyperlinks are creating a new type of thinking—part human and part machine—found nowhere else on the planet or in history."[21] Such a notion of networked, machine-aided thought seems barely removed from Wells's aspirations for a "greater mental superstructure," Otlet's "mechanical and collective brain," or Engelbart's hopes for "augmenting human intellect."

That notion of an emergent, bottom-up system has also fueled speculation about the possibility of so-called collective intelligence, the notion that new forms of thought may emerge out of the ether of collaborative intellectual work taking place on the Web. The idea of distributed, networked intelligence has a long pedigree, as we have seen. Otlet, Wells, and de Chardin all imagined a new kind of planetary consciousness springing forth out of the network. More recently, the French philosopher Pierre Lévy has articulated a similar hope for a humanistic knowledge-sharing environment, writing of the possibility of a new collective intelligence emerging, which he defines as "a form of universally distributed intelligence, constantly enhanced, coordinated in real time, and resulting in the effective mobilization of skills." To realize that vision, he envisions new, networked information systems that will allow large numbers of individuals and groups of experts to collaborate, contributing to "the same virtual universe of knowledge." And, like Otlet, Lévy hopes that such an environment will have far-reaching benefits beyond simply making it easier to share information online. "The ideal of collective intelligence implies the technical, economic, legal, and human enhancement of a universally distributed intelligence."[22]

Behind so much of the discussion lies Otlet. I have noted the particular examples of overlap and separation between his projects and those of Nelson, Berners-Lee, and others, but what is striking is the degree to which Otlet framed the debate. He never framed his thinking in purely technological terms; he saw the need for a whole-system approach that encompassed not just a technical solution for sharing documents and a classification system to bind them together, but also the attendant political, organizational, and financial structures that would make such an effort sustainable in the long term. And while his highly centralized, controlled approach may have smacked of nineteenth-century cultural imperialism (or, to put it more generously, at least the trappings of positivism), it had the considerable advantages of any controlled system, or what today we might call a "walled garden": namely, the ability to control what goes in and out, to curate the experience, and to exert a level of quality control on the contents that are exchanged within the system.

Today's Web stands as both a vindication of the early hypertext seers as well as a useful case study in one of the technology industry's core axioms: that the best technology does not always win. The early history of hypertext points toward a number of intriguing roads not taken (Bush's two-way linking, Nelson's transclusion, and Otlet's auxiliary tables, to name a few). While the absence of these features certainly hasn't stopped the Web from rising to its current heights of popularity—there are currently more than 600 million Web sites in existence[23]—at the same time, it is worth noting that more than a few successful sites have profited handsomely by filling the void left by the Web's lowest-common-denominator approach. For example, Facebook's policy of using real names, coupled with its vast user base, has helped it achieve a role as institutional guarantor of personal identity online. Users entrust the site with their personal information (Facebook's occasional privacy policy kerfuffles notwithstanding), and in turn Facebook provides them with a reasonable level of trust that they can interact with friends and associates, exchanging sometimes intimate conversations and photos and sharing links to documents elsewhere on the Web. Facebook has so established itself as an arbiter of identity that many third-party sites now rely on its Facebook Connect program to support login to their own sites and applications, to avoid maintaining their own registration databases. The lack of reliable identity management has bedeviled the Web from the beginning, a problem that stems in large part from the fundamentally distributed design of the Internet, which prevents the establishment of such central authorities for the network as a whole. H. G. Wells recognized the importance of stable identity management early on and made it one of the conceptual linchpins of his World Brain. Absent such controls on the Web, companies like Facebook now play this role de facto for a large percentage of the Web's population.

Facebook also employs linking mechanisms that closely resemble the bidirectional links proposed by Bush, Nelson, and so on. Whereas a typical Web hyperlink usually goes in one direction—from source document to target hyperlink—links on Facebook typically work both ways; a comment or "like" interaction will show up in multiple threads; and a user may follow a link back to see the original commenter's page. Bush also envisioned the possibility of "visible trails," in which a given user could follow another user through an information space. The Twitter community's now-prevalent use of hashtags (the now-ubiquitous # symbol preceding a keyword to concatenate posts about the same subject, as in "#thewebthatwasnt") seems to serve much the same purpose, allowing for a user to pull together numerous posts from different users commenting on the same topic or link.

For all their foresight about the possibilities of information networks, none of the hypertext pioneers truly saw the social media revolution coming. Otlet spoke of allowing individual users to "sing in the chorus," an early if cautious form of collective filtering. But he certainly never envisioned them writing the music as well. Similarly, neither Bush nor Nelson envisioned networked environments designed primarily for individual consumption, rather than group interaction. And while Engelbart thought deeply about creating tools to help groups work together, he had in mind small teams in the workplace rather than an environment to support casual social interactions on a global scale.

As much as the Web has fulfilled the vision of the early pioneers as a global knowledge repository, in day-to-day use it looks less like Otlet's library than a bazaar, or perhaps an unruly town square. Whereas in its early days the Web did seem to function as the kind of distributed research tool that Otlet, Bush, and the rest imagined, by the mid-2000s a new reality seemed to be coming into focus. The explosive rise of social media companies like Facebook, Twitter, Instagram, and others has been recounted ad nauseam in the press and scarcely needs recapping here. But it is worth noting that the emergence of the social sphere altered the interfaces that most users now rely on to create and retrieve information. While the early Web consisted primarily of static documents, today's real-time Internet gives increasing prominence to the "stream" metaphor now so prevalent in social media, in which a chronological display of information, carefully filtered through networks of association, seems gradually to be displacing the more structured, static interfaces of the Web's early years.

In years to come, the Web will likely continue to evolve from a far-flung agglomeration of unrelated sites to what computer science pioneer David Gelernter has called "the worldstream," which he describes as an online environment in which "billions of users will spin their own tales, which will merge seamlessly into an ongoing, endless narrative." While the future he imagines may differ from Otlet's in the mechanics of its interactions, it shares the same spirit of enlightenment. "No one can see the whole worldstream, because much of the information flowing through it is private. But everyone can see part of it," he writes[24]—seeming to echo Otlet's notion of a personalized knowledge system that would let anyone tap into humanity's entire intellectual output, constructing a personalized information retrieval system from the comfort of one's armchair.

Otlet's thinking about the evolutionary progression of human knowledge and the possibility of ultimate transcendence also finds strong echoes in contemporary literature about the Internet. While the trippy liberation ethos—one might even say theology—of early visionaries like Stewart Brand and Ted Nelson may seem a far cry from the orderly, World Government orientation of Otlet and his fellow belle epoque thinkers, they share a common belief in utopian ideals fueled by free-flowing access to information and the breaking down of traditional structures of knowledge. The contemporary construct of "the user" that underlies so much software design figures nowhere in Otlet's work.[25] He saw the mission of the Mundaneum as benefiting humanity as a whole, rather than serving the whims of individuals. While he imagined personalized workstations (those Mondotheques), he never envisioned the network along the lines of a client-server "architecture" (a term that would not come into being for another two decades). Instead, each machine would act as a kind of "dumb" terminal, fetching and displaying material stored in a central location.

The counterculture programmers who paved the way for the Web believed they were participating in a process of personal liberation. Otlet saw it as a collective undertaking, one dedicated to a higher purpose than mere personal gratification. And while he might well have been flummoxed by the anything-goes ethos of present-day social networking sites like Facebook or Twitter, he also imagined a system that allowed groups of individuals to take part in collaborative experiences like lectures, opera performances, or scholarly meetings, where they might "applaud" or "give ovations." It seems a short conceptual hop from here to Facebook's ubiquitous "Like" button.[26]

The notion of a "world brain" once evoked by H. G. Wells and Otlet has found plenty of adherents in the modern era. Contemporary pundits like Ray Kurzweil, Howard Bloom, Kevin Kelly, and others have all advocated the possibility of a global planetary awakening, as the Web takes us to the next step in the evolution of human consciousness. In the end, what distinguishes Otlet's vision from these cyber-utopians is his belief in the positive role of institutions. More than simply individuals were enlightened; institutions too could be enlightened. Otlet saw his ideal society as a perpetual work in progress, one that would require constant effort and adaptation—an aspirational ideal more than an ultimate end state. He saw human culture as a living organism, forever moving forward on the never-ending path to realization.

Otlet's notion of evolutionary progress stemmed of course from his positivist beliefs, which were unshakeable, however much his dreams were disappointed. Contemporary evolutionary biologists tend to regard claims for cultural "evolution" with a degree of skepticism. That said, his invocation of an evolutionary paradigm for the propagation of networked information anticipates a stream of rhetoric that characterized the early years of the Web. In a 1998 article for *Wired* magazine, programmer Danny Hillis described the emergence of the contemporary global network in explicitly evolutionary terms. "We're taking off," he wrote. "We are not evolution's ultimate product. There's something coming after us, and I imagine it is something wonderful."[27] Kurzweil has suggested that advances in information technology have led humanity to the brink of an evolutionary event: the Singularity, a point at which we will witness "the merger of biological and nonbiological intelligence."[28]

Otlet's notion of "collective intelligence" prefigured a great deal of contemporary pro-Internet rhetoric. The publisher Tim O'Reilly has embraced the same term; and social media acolytes like Clay Shirky have celebrated the ability of the global network to magnetize millions of connected minds.[29] While Internet pundits have celebrated this evident facility for shared knowledge generation as one of the great boons of the modern networked age, the notion stretches back to Émile Durkheim, whose writings on collective consciousness formed an important backdrop for Otlet's treatment of the subject in *Monde*. But whereas Durkheim conceived of collective intelligence in terms of social interaction between individuals, Otlet took the concept far more literally. To realize the power of collective intellect it would be necessary to build the instruments and institutions to facilitate it: The Universal Bibliography, the Union of International Associations, and the World City would all play critical roles in helping humanity tap its greater potential for collective thinking and problem solving. For Otlet, these developments would mark an important step in the evolution of the species or, depending on one's point of view, humanity's quest for salvation. Many have come to view evolutionary explanations of cultural phenomena as highly problematic (witness the sociobiology wars of the 1970s that engulfed E. O. Wilson's attempts to explain cultural progress through evolutionary mechanisms). Whether we take the notion of cultural "evolution" as a literal biological process or just a useful metaphor, the notion that the Web represents a great leap forward for humankind seems to command no shortage of adherents.

Otlet's work invites us to consider a simple question: whether the path to liberation requires maximum personal freedom of the kind that characterizes today's anything-goes Internet, or whether humanity would find itself better served by pursuing liberation through the exertion of discipline.

Conclusion

On an unseasonably warm March morning in 2012, Belgian prime minister Elio di Rupo joined a group of scholars, European Union representatives, and assorted hangers-on at Google's Brussels office near the Parc Leopold—a few steps from the former site of Otlet's Palais Mondial—to announce a new partnership between Google and a tiny museum in Mons, a French-speaking city in the province of Hainaut, called the Mundaneum.

In preparation for the event, the Google public relations team furnished a slickly produced multimedia video recounting the story of Paul Otlet and celebrating his role as a forerunner of the Internet. The Google partnership came as a welcome boost for the museum, whose archivists have been working since 1998 to preserve the legacy of Otlet and La Fontaine. In physical form, that legacy is now stored in thousands of cardboard boxes (Otlet's papers alone take up more than 1,000 boxes, most of them still awaiting cataloging). The staff is in the process of converting Otlet's archives to a digital repository, to be made available over the Web. Along the way, they have also hosted a series of exhibits, scholarly colloquia, and cultural outreach programs designed to keep the spirit of the Mundaneum alive.

Reproduction of the Mondotheque (2012), based on Otlet's original design. Reproduced with permission from the Mundaneum.

Ten years before, almost no one outside of specialized academic circles had heard of Paul Otlet. Now he was finally getting his due as one of the pioneers of the Information Age. For the museum, the Google partnership meant new publicity (and funding) for their work. But what did Google gain? A cynic might attribute their motives to pure public relations. Google's recently established Brussels office—charged with managing public relations for the search giant's contentious relationship with the European Union over privacy issues—seemed eager to prove its cultural bona fides and establish itself as a good corporate citizen. Indeed, Google has since initiated a series of other cultural sponsorships across Europe, trying to project an image in keeping with its corporate motto: "Don't Be Evil," a mantra that seems a distant echo of Otlet's dictum that the Mundaneum must "always be good."

While Google's troubled relationship with the European Union may have provided some impetus for the sponsorship, there were human factors at work as well. Google's local marketing manager, Julien Blanchez, happened to hail from Mons, home of the Mundaneum museum, as did Prime Minister di Rupo, who for a time served as the city's mayor. In this case, local pride may have played at least as much of a role as corporate self-interest.[1] Whatever the case, the Mundaneum project fit well with Google's organizational rhetoric, which often seems to project a positivist, even utopian, ethos in its public proclamations. Otlet himself would surely have embraced their stated mission: "to gather the world's information."

In 2008, Google founder Larry Page helped launch the Singularity University, a gathering of technologists dedicated to preparing for a Wellsian moment of technological transcendence predicted by, among others, Ray Kurzweil. It would take shape as a new form of collective intellect over the network as human minds became increasingly fused with so-called thinking machines within a vast global network. Google has often laid claim to a higher social purpose. In 2004, technology historian George Dyson reflected on his visit to the Googleplex in a much-discussed essay entitled "Turing's Cathedral": "I felt," wrote Dyson, "I was entering a 14th-century cathedral—not in the 14th century but in the 12th century, while it was being built. Everyone was busy carving one stone here and another stone there, with some invisible architect getting everything to fit. The mood was playful, yet there was a palpable reverence in the air."[2] That spirit of reverence underlies much of what Google has tried to accomplish—not just delivering search results, but creating self-driving cars, space elevators, humanitarian tools for crisis response, and a scientific platform for environmental forecasting, as well as any number of unknown classified projects going on at any given time in the company's secretive X Lab, under the close supervision of co-founder Sergey Brin. But many contemporary critics have questioned how such a lofty-sounding mission can square with the commercial imperatives of a publicly traded, for-profit enterprise.

For all its rhetoric and farsighted initiatives, the company nonetheless has always made its money as would an old-fashioned media company: by selling advertising. That gushing revenue stream has given Google the luxury of pointing some of its smartest engineers at long-term problems. Nonetheless, the ineluctable logic of capitalism dictates that the company must generate a profit; and that tension between "gathering the world's information" and promoting its own products and services has brought Google into conflict with regulators the world over.

Today, Google and a handful of other major Internet corporations like Facebook, Twitter, and Amazon play in effect much the same role that Otlet envisioned for the Mundaneum—as the gathering and distribution channels for the world's intellectual output. But the parallels only run so far. Commercial enterprise played at best a minor role in the world that Otlet envisioned. He envisioned a publicly funded, transnational organization—not a tournament of for-profit companies. He likely would have seen the pandemonium of today's Web as an enormous waste of humanity's intellectual and spiritual potential.

Whether such an organizational model would ever have worked in practice is anyone's guess. In 2008, the European Union announced its backing for Quaero, an ambitious but ultimately ill-fated search engine project designed to create a competitor to Google. Launched with much fanfare by French president Jacques Chirac in 2006, the pan-European initiative enlisted a team composed primarily of French and German developers. Chirac compared the effort to Airbus, a similar pan-European government-funded effort to boost the continent's lagging participation in the burgeoning aeronautics industry. The organization set out to create a viable European competitor to the American search giants Google and (then) Yahoo! But the project soon fell apart, as the French and German teams failed to come to consensus on their priorities (the Germans wanted to build a text-based search engine; the French preferred to focus on the multimedia search, which they considered a more interesting and largely unsolved problem). The German team eventually withdrew from the project, and the organizers pared back their ambitions. The company has since changed its name to Theseus, producing a scaled-down product that has met with limited success.

Meanwhile, the Google office now sits in a brightly lit, glass-walled space near the European Quarter in Brussels, where employees focus primarily on advertising, marketing, and hosting the company's central communications office for Europe—and enjoying the company's famously free lunch in a well-appointed cafeteria. Outside the Google building, citizens from across Europe go to work in the sleek-looking office buildings of the European Union and its affiliated organizations. Looking at them, it's hard not to be reminded of various versions of the World City, and what Otlet, Andersen, and Le Corbusier had once imagined. Beyond the immediate sphere of the European Union, Belgium also plays host to a shadow industry of contractors, lobbyists, and public relations units. It is no accident that Google chose to centralize its pan-European public relations operation in Brussels—or invest in the person of William Echikson, a former Brussels bureau chief for Dow Jones who for the past five years has headed up Google's efforts involving the European Union. He has also served as the point person for Google's relationship with the museum in Mons. The Google staffers have even christened one of their conference rooms "Mundaneum."

The real—if that's the right word for it—Mundaneum sits in a former department store and onetime parking garage on a quiet street off the town square in Mons, a medieval city better known to history as the site of the first battle between British and German troops during World War I. Today, the city hosts a university as well as nearby offices of Google and Microsoft. The location of the Mundaneum in Mons—a city that otherwise claims no particular connection to Paul Otlet—stems from a personal connection with its former mayor, Prime Minister Di Rupo. Established in 1996, the museum features an Art Deco–style exhibit space designed by noted Belgian comic artists François Schuiten and Benoît Peeters, and a separate, more dour-looking archival facility. Today, a staff of researchers and archivists labor there in relative seclusion, trying to reconstruct a version of the original Mundaneum, in hopes of bringing Otlet's vision to life. With the help of Google funding as well as ongoing state support, the museum also hosts public exhibits and lectures designed to encourage civic discussion. Most recently they have been running a series of Web-based videos and lectures about European forerunners of the Internet. As of this writing the museum is temporarily closed for major renovations in anticipation of the European Union's planned designation of Mons as the 2015 European Capital of Culture.

The Mundaneum houses much of the archival material used in this book, including Otlet's diaries, many of the original file cabinets and index cards that comprised the Universal Bibliography, and a vast assembly of boxes containing books, pamphlets, photographs, posters, and all manner of other ephemera. These have supported a cottage industry of scholarship on Otlet that has sprung up in the academic world over the past few years, primarily in the domains of information science and documentation, but increasingly in such other fields as architecture and art history. The Otlet revival owes principally to the efforts of Boyd Rayward, the Otlet scholar who almost single-handedly recovered his legacy in 1968 and who has since devoted a good part of his career to studying Otlet's work. Rayward is currently at work on a new study of Otlet, one that will likely be both authoritative and encyclopedic.

In Rome, the Hendrik Andersen Museum occupies a building on a sleepy street on the edge of the old city, a few blocks from the majestic Piazza del Popolo. Occupied by Andersen during the latter stages of his life, it was bequeathed to the Italian government by the Andersen estate and is now managed by the Italian Ministry of Culture. The building displays some of his sculptures that had been intended for the World City and wall-sized prints of Ernest Hébrard's architectural drawings, as well as a few smaller, more intimate works, including busts of Henry James and Andersen's sister-in-law and lifelong supporter, Olivia Cushing.

When the institution opened to the public in 2000, the *New York Times* dubbed it a "bizarre museum" and "a monument to a talent that went off the rails."[3] Indeed, the oversized sculptures—which had made James cringe—seem today like cartoonish parodies of the classical (or socialist) ideal. Heroic-looking nude men, women, and children strike exaggerated poses while staring into some distant utopian future. That stout spirit seems particularly out of place in this particular museum, administered by a staff that seems a little startled at the arrival of an actual visitor. The place has the feel of an heirloom handed down from some distant relative that no one really wants but feels an obligation to maintain.

In downtown Edinburgh, the Outlook Tower remains a popular attraction, and visitors still come to marvel at Patrick Geddes's old Camera Obscura, whose real-time streaming images from the street seem no less arresting in their verisimilitude even in an age of traffic cams and YouTube. While the camera remains intact, Geddes's Index Museum has given way to a showy multimedia exhibit called the World of Illusions, featuring a magic gallery, light show, and assorted optical illusions. A faint echo of Geddes's spirit seems to live on, however, in an exhibit titled "Edinburgh Vision," which provides a historical perspective on the city's evolution over the past 150 years through a combination of stereoscopic lenses, mirrors, and 3-D glasses.

Elsewhere, Geddes's legacy as an influential sociologist and town planner seems secure. His ideas about urban planning, rooted in first-hand observational research, went on to influence generations of important thinkers, including Lewis Mumford, who once praised him as a "global thinker in practice, a whole generation or more before the Western democracies fought a global war."[4] Geddes's work continues to suggest intriguing possibilities for rethinking the architecture of information: creating physical spaces that complement and extend our ability to access large bodies of networked information.

While the international museums Geddes, Otlet, and Neurath envisioned never came about in quite the way they imagined, the spirit of that global museum has recently found expression in China, where the Window of the World museum in Shenzhen, China—Deng Xiaoping's ready-made megalopolis—hosts thousands of visitors who come to see the world's wonders reproduced in miniature in this unabashedly internationalist museum. A 1:3-scale replica of the Eiffel Tower stands alongside similarly diminutive doppelgangers of the Lincoln Memorial, which sits side-by-side with reconstructions of Angkor Wat, the Taj Mahal, and other landmarks. "Placing all the world's architectural wonders on an equal platform breaks down the distinctions between peoples," writes journalist Daniel Brook, "and inspires in visitors a cosmopolitan, human pride."[5] Such a sentiment embraces internationalist enthusiasm that inspired Geddes's Index Museum, or the later attempts at syndicated museum exhibits by Otlet and Neurath. Otlet's World Palace of course came in for its share of criticism for its conceptual overreach and seemingly haphazard collection development policies. So too the Shenzhen museum, as Brooks puts it, "embodies contemporary China at its tawdriest."

A few years after Otlet's death in 1944, his former collaborator Le Corbusier led a team of architects in designing the new United Nations headquarters in New York. During the planning process, he frequently referred to his earlier work with Otlet on the World City. In his reflections on the project in 1957, he once again invoked the spirit of the old friend he used to call Saint Paul. "Like a Wandering Jew," he wrote, "this World City seeks a home."[6]

The spirit of internationalism that drove Otlet has not disappeared. From the United Nations and its constellation of affiliated nongovernmental organizations to the steady stream of news articles chronicling the effects of globalization, the evidence for it is everywhere. Thomas L. Friedman and others gave a popular voice to the argument that the Internet and its protocols, coupled with the rise of open-source software, have provided a "crude foundation of a whole new global platform for collaboration."[7] In this networked world, traditional hierarchies—national, organizational, and ontological—are giving way to a "flat" world, characterized by long-distance collaboration, offshoring, and rapid advances in computing speed and communication protocols. Meanwhile, the world's cultures seem increasingly to converge around shared reference points: movies, music, television shows, fashion, and dance videos that seem to be coalescing into a kind of cultural lingua franca.

While all this global awareness bears a superficial resemblance to Otlet's internationalism, it also seems a long way from his vision of a transglobal government administering a universal knowledge network for the benefit of humanity. Otlet's faith in the beneficent power of associations and institutions seems anachronistic in our contemporary decentered networked age, when the Internet has proven its ability to disrupt—rather than unite—traditional organizational structures. And the corporate imperatives that drive so much innovation on the Internet—Google's high-minded proclamations notwithstanding—seem far removed from the quest for universal truth and spiritual transcendence that motivated Otlet.

Would the Internet have turned out any differently had Paul Otlet's vision come to fruition? Counterfactual history is a fool's game, but it is perhaps worth considering a few possible lessons from the Mundaneum. First and foremost, Otlet acted not out of a desire to make money—something he never succeeded at doing—but out of sheer idealism. His was a quest for universal knowledge, world peace, and progress for humanity as a whole. The Mundaneum was to remain, as he said, "pure." While many entrepreneurs vow to "change the world" in one way or another, the high-tech industry's particular brand of utopianism almost always carries with it an underlying strain of free-market ideology: a preference for private enterprise over central planning and a distrust of large organizational structures. This faith in the power of "bottom-up" initiatives has long been a hallmark of Silicon Valley culture, and one that all but precludes the possibility of a large-scale knowledge network emanating from anywhere but the private sector.

Otlet's Mundaneum would also have differed from today's Web in several of its core functions. Setting aside its technological limitations—like storing data on index cards—the Mundaneum envisioned a far more sophisticated mechanism for cross-referencing information between sources, tracking versions, and fostering collaboration among users. Such a high degree of coordination and classification standards remains, as we've seen, conspicuously absent from the Web. While the Semantic Web community has explored the possibility of more structured ontologies, the sheer scope and size of the Web, coupled with its fundamentally distributed architecture, seem to preclude the possibility that any one entity will ever manage the whole thing.

And that is undoubtedly as it should be; the triumph of the Web owes a great deal to its fundamental disorder. That said, a few control mechanisms might open up intriguing possibilities: better intellectual property protections; the ability to encode more sophisticated kinds of hyperlinks (such as denoting whether or not a particular document "agrees" with another one); and the ability to cross-reference and collate material in such a way as to support better synthesis—in the form of summary documents, visual indexing tools, or other "zoomable" interfaces that would allow users to drill down into accessible levels of detail. Such sophisticated interactions could yield enormous benefits in certain focused areas of inquiry—medicine, the sciences, or law, for example—where researchers would benefit from a more tightly managed system that could give them comprehensive access to published literature in a more efficient manner. Indeed, many specialty publishers and database providers—such as MEDLINE, the Web of Science, and Lexis/Nexis—continue to generate profits by serving these specialized communities. And the Semantic Web community sees enormous potential in creating so-called Webs of Data to serve these and other specialized communities. While it seems unlikely that Otlet's classification scheme will ever supplant any of these initiatives, his vision does offer an organizational model for world-wide knowledge sharing, fostered through an enterprise devoid of corporate profit motives.

That is why Paul Otlet still matters. His vision was not just cloud castles and Utopian scheming and positivist cant but in some ways more relevant and realizable now than at any point in history. To be sure, some of his most cherished ideas seem anachronistic by today's standards: his quest for "universal" truth, his faith in international organizations, and his conviction in the inexorable progress of humanity. But as more and more of us rely on the Internet to conduct our everyday lives, we are also beginning to discover the dark side of such extreme decentralization. The hopeful rhetoric of the early years of the Internet revolution has given way to the realization that we may be entering a state of permanent cultural amnesia, in which the sheer fluidity of the Web makes it difficult to keep our bearings. Along the way, many of us have also entrusted our most valued personal data—letters, photographs, films, and all kinds of other intellectual artifacts—to a handful of corporations who are ultimately beholden not to serving humanity but to meeting Wall Street quarterly earnings estimates. For all the utopian Silicon Valley rhetoric about changing the world, the technology industry seems to have little appetite for long-term thinking beyond its immediate parochial interests.

Unlike Otlet's patron Andrew Carnegie—who devoted the latter part of his life to giving away his fortune in support of public-minded causes like peace conferences, public libraries, and the Palais Mondial—today's Internet billionaires often seem concerned primarily with their own self-glorification: building spaceships, ultra-fast yachts, or staging multimillion-dollar weddings. (Bill Gates is so far the exception that proves the rule; his personal foundation has made impressive strides toward lowering childhood mortality rates worldwide, embracing the Carnegie ethos of dispensing a great fortune for the public good.[8]) The fundamentally libertarian ethos of Silicon Valley, coupled with the decentralized systems architecture of the Internet, seems to preclude the possibility that anyone could now hope to build the kind of global intellectual nerve center that Otlet imagined.

Paul Otlet with model of the World City (1943). Reproduced with permission from the Mundaneum.

Otlet's Mundaneum will never be. But it nonetheless offers us a kind of Platonic object, evoking the possibility of a technological future driven not by greed and vanity, but by a yearning for truth, a commitment to social change, and a belief in the possibility of spiritual liberation. Otlet's vision for an international knowledge network—always far more expansive than a mere information retrieval tool—points toward a more purposeful vision of what the global network could yet become. And while history may judge Otlet a relic from another time, he also offers us an example of a man driven by a sense of noble purpose, who remained sure in his convictions and unbowed by failure, and whose deep insights about the structure of human knowledge allowed him to peer far into the future.

Whether or not the Internet proves to be a good thing for humanity is another question. At the very least its impact on all of our lives requires us continually to think ahead and at the same time to look to the past for guidance. Otlet, somewhere both behind us and ahead of us, is a worthy avatar. His work points to a deeply optimistic vision of the future: one in which the world's knowledge coalesces into a unified whole, narrow national interests give way to the pursuit of humanity's greater good, and we all work together toward building an enlightened society.

ACKNOWLEDGMENTS

All books are to some extent made out of the books that came before, as Otlet knew full well. And so, in that Otletian spirit, let me begin with a nod to the many works cited in the preceding pages, and to the authors, editors, librarians, and assorted "bibliologists" who made it possible for me to consult them. My name may appear on the cover, but this work is in truth the product of many minds.

This book likely would not exist at all if it were not for the seminal work of Boyd Rayward, whose years of careful research and diligent scholarship have brought Otlet's oeuvre into contemporary focus. I have had the good fortune to get to know Boyd personally over the past few years, and have found him unfailingly generous with his time and expertise; he is the living embodiment of that well-worn phrase, "a gentleman and a scholar." I also owe a debt of gratitude to the wider community of Otlet scholars, especially Charles van den Heuvel and Wouter van Acker, both of whom offered critical feedback on portions of the manuscript. The staff of the present-day Mundaneum, especially Stéphanie Manfroid, Delphine Jenart, and Raphaèle Cornille, also provided invaluable support and encouragement along the way.

My editor, Timothy Bent, took a chance on this book and proceeded to steer its trajectory with a rare degree of skill; his storytelling instincts and laser eye for detail have made this book immeasurably better than it would otherwise have been; Keely Latcham also played a key role in shepherding the manuscript into being. Rich Meislin and Laura Chang at the *New York Times* helped shape the November 2008 article that provided the initial spark for this book. My agent, Laurie Liss, saw the potential in that article for a book, and it is largely thanks to her gentle prodding that I embarked on this project in the first place.

My teacher and friend Charles Strozier offered insightful critiques of my early drafts; his perspective on narrative nonfiction and the uses of historical context have deeply influenced the direction of this book. My dear friend Mary Ann Caws volunteered her time and expertise in poring over Otlet's often-indecipherable diary entries; I felt blessed to have such a masterful translator at my side. Julia Toaspern provided skillful translations of German source materials. Thanks also to the friends and colleagues who reviewed the manuscript at various stages of completion, especially my *vieux amis* Whit Andrews and Mike Myers. Other readers included Jeffrey Colvin, Liz Danzico, Janet Hadda, Laura Hoopes, Sangamithra Iyer, Michael Keane, Toni Logan, and Wayne Soini.

For institutional support, I am grateful to the Mundaneum, the New York Public Library, the Royal Library of Belgium, the Library of Congress Manuscript Division, the Norman Mailer Writers Colony, and the Brooklyn Writers Space.

Finally, my deepest thanks go to my family: to my wife, Maaike, for her love and forbearance; and to my sons, Colin and Elliot, who continually inspire me with their living examples of spontaneous wit and wisdom.

NOTES

Introduction

1. "Documentation Congress Step toward Making 'World Brain.'"
2. Unknown, "Einsatzstab Reichsleiter Rosenberg [papers]," f. 3676.
3. Simon, "Chronologie Krüss, Hugo Andres."
4. Kurtz, *America and the Return of Nazi Contraband*, 22–25.
5. Wright, *Glut*, 54–93.
6. Unknown, "Einsatzstab Reichsleiter Rosenberg [papers]," f. 3676.
7. Otlet, *Traité de documentation*, 238.
8. Was Otlet's "electric telescope" really a computer? If we take the word "computer," in the Merriam-Webster sense, to mean a "programmable electronic device," then certainly not. But if we take the Wittgensteinian view that a word is defined by its use, then Otlet's "electric telescope" does indeed seem to resemble what most people would consider a computer: that is, a connected device for retrieving information over a network.
9. Van den Heuvel, Paul Otlet et les versions historiques de la genèse du World Wide Web, du Web Semantique et du Web 2.0
10. Otlet, *Monde: Essai d'universalisme*, 391.
11. Otlet, as quoted in Levie, *L'Homme qui voulait classer le monde*, 291. Translation mine.
12. Ibid., 287.
13. Note that it is not clear whether Otlet ever actually sent this letter (per Boyd Rayward, personal conversation).
14. Unknown, "Einsatzstab Reichsleiter Rosenberg [papers]," f. 3676.
15. Rayward, La Fontaine, and Otlet, *Mundaneum: Archives of Knowledge*.

16. Manfroid and Jenart, interview with the author.
17. Rayward, "The Case of Paul Otlet."
18. Ibid.
19. Borges, "Library of Babel."
20. Kelly, "We Are the Web."
21. Pritchard, "Risk of Information Overload That Threatens Business Growth."
22. Wright, "The Web Time Forgot"; Laaff, "Internet Visionary Paul Otlet."

Chapter 1

1. Gessner, as quoted (translated) in Krajewski, *Paper Machines*, 13.
2. Blair, *Too Much to Know*, 214–17.
3. Gessner, *Bibliotheca Universalis Und Appendix.*
4. Library of Congress, "Fascinating Facts."
5. Febvre and Martin, *The Coming of the Book*, 248.
6. Blair, *Too Much to Know.*
7. Eisenstein, *Printing Revolution in Early Modern Europe*, 65.
8. Ibid., 69.
9. Columbanus, *Monk's Rules*; cited in Greenblatt, *The Swerve*, 27–28.
10. Erasmus and Barker, *The Adages of Erasmus*, 145.
11. Calvin, as quoted (translated) in Blair, *Too Much to Know*, 56.
12. Gessner, as quoted (translated) in ibid.
13. Descartes, as quoted (translated) in ibid., 5.
14. Foxe, *Unabridged Acts and Monuments Online*, 858 (original spelling modernized for clarity).
15. Wall Randell, "Dr. Faustus and the Printer's Devil," 259–81.
16. Bacon, *Sylva Sylvarum.*
17. Diderot, *Oeuvres de Denis Diderot.*
18. Eco, *The Search for the Perfect Language*, 246.
19. Leibniz, as quoted (translated) in *World of Mathematics*, 64.
20. Dyson, *Darwin among the Machines*, 7.
21. Leibniz, as quoted (translated) in Blair, *Too Much to Know*, 58.
22. Ibid., 228.
23. Krajewski, *Paper Machines*, 33–35.
24. Taylor, *Principles of Scientific Management*, 7.
25. Buckland, "On the Cultural and Intellectual Context of European Documentation," 46.
26. Wiegand, *Irrepressible Reformer*, 22.
27. Dewey, "[Editorial]."
28. Dewey, *Classification and Subject Index.*
29. Weinberger, *Everything Is Miscellaneous*, 56.

30. Dewey, *Classification and Subject Index.*
31. Weinberger, *Everything Is Miscellaneous,* 57.
32. Dewey, *Classification and Subject Index.*
33. Krajewski, *Paper Machines,* 3.
34. Ibid., 65.

Chapter 2

1. Emerson, *Leopold II of the Belgians,* 24–25.
2. "Industrial History | Belgium."
3. Otlet, as quoted in Levie, *L'Homme qui voulait classer le monde,* 17. Translation mine.
4. Rayward, *Universe of Information,* 10.
5. Otlet, "Journal Intime," November 2, 1885.
6. Ibid., October 4, 1885.
7. Ibid., November 2, 1885.
8. Otlet, as quoted in Levie, *L'Homme qui voulait classer le monde,* 32. Translation mine.
9. Otlet, "Journal Intime," January 1, 1883.
10. Borges, "Analytical Language of John Wilkins."
11. Otlet, as quoted (translated) in Rayward, *Universe of Information,* 18.
12. Ibid., 17.
13. Otlet, "Journal Intime."
14. Ibid.
15. Leopold II, as quoted (translated) in Hochschild, *King Leopold's Ghost,* 38.
16. Ibid., 159.
17. King Leopold II, quoted in ibid., 44.
18. Ibid., 58.
19. Ibid.
20. Twain, *King Leopold's Soliloquy.*
21. Brooker et al., *Oxford Critical and Cultural History of Modernist Magazines,* 337.
22. Hochschild, *King Leopold's Ghost,* 4.
23. Leopold II, quoted (translated) in Emerson, *Leopold II of the Belgians,* 241.
24. Ibid., 241.
25. Otlet, *L'Afrique aux noirs,* 13–14.
26. Ibid., 10–11.
27. Otlet, *Leopold II et nos villes.*
28. Rayward, "Knowledge Organisation and a New World Polity," 5.
29. Levie, *L'Homme qui voulait classer le monde,* 39.
30. Comte, as quoted (translated) in Sklair, *Sociology of Progress,* 39.
31. Otlet, as quoted (translated) in Rayward, *Universe of Information,* 20.

32. Otlet, as quoted (translated) in Van Acker, *Universalism as Utopia*, 79.
33. Otlet, as quoted (translated) in Rayward, *Universe of Information*, 38.
34. Ibid., 24.

Chapter 3

1. Boyd, *Paris Exposition of 1900*, 141.
2. Willems, preface to Robida, *Twentieth Century*.
3. Watson, *Ideas*, 729.
4. Steenson, "Interfacing with the Subterranean."
5. Williams, *Notes on the Underground*; cited in ibid.
6. Mazower, *Governing the World*, 103–6.
7. Watson, *Ideas*, 729.
8. [Unnamed], as quoted in Mazower, *Governing the World*, 106.
9. Mazower, *Governing the World*, 106.
10. U.S. Commission to the Paris Exposition and Peck, *Report of the Commissioner-General for the United States to the International Universal Exposition*, 377–79.
11. Ryan, "About ESW and the Holocaust Museum."
12. Jacobs, *Death and Life of Great American Cities*, 15.
13. Delombre, as quoted (translated) in Boyd, *Paris Exposition of 1900*, 107.
14. Delombre, as quoted (translated) in ibid., 107.
15. Ibid., 112.
16. *Henri La Fontaine, Prix Nobel de La Paix en 1913*, 9.
17. Comte, as quoted (translated) in Sklair, *Sociology of Progress*, 39.
18. A new typewriter sold for approximately $100 in 1897, or the equivalent of about $2,500 in today's currency.
19. Rayward, *Universe of Information*, 62.
20. Henri Stein, as quoted (translated) in ibid., 59.
21. Given the difficulty of distinguishing the two organizations' often overlapping activities, they are hereafter referred to as the IIB, except in those cases where the OIB was engaged in specific initiatives related to the Belgian government.
22. Schollaert, as quoted in Levie, *L'Homme qui voulait classer le monde*, 59. Translation mine.
23. La Fontaine, as quoted (translated) in Rayward, *Universe of Information*, 34.
24. Ibid., 120–21.
25. Richet, as quoted (translated) in ibid., 64.
26. Rayward, "Paul Otlet. Encyclopédiste, Internationaliste, Belge."
27. Rayward, *Universe of Information*, 86.
28. Otlet, as quoted (translated) in Rayward, *Universe of Information*, 113.
29. Otlet, "Something about Bibliography," 11.
30. Ibid., 17.
31. Eco, "Vegetal and Mineral Memory."

32. Otlet, "Something about Bibliography," 17.
33. Ibid., 20.
34. Otlet, "Sur la structure des nombres classificateurs," 230–43.
35. Hunter, *Classification Made Simple*, 65–68.
36. "UDC Structure & Tables."
37. Otlet, "Transformations of the Bibliographical Apparatus of the Sciences," 153.
38. Rayward, *Universe of Information*, 75.
39. Otlet, as quoted (translated) in Van Acker, *Universalism as Utopia*, 96.

Chapter 4

1. Postman, *Amusing Ourselves to Death*, 67.
2. Standage, *Victorian Internet*.
3. Samuel Morse, letter to C. T. Jackson, September 18, 1837, as quoted in Batchen, "Electricity Made Visible," 36–37.
4. Reef, *Walt Whitman*, 32.
5. Whitman, *Leaves of Grass*.
6. Twain, "From the 'London Times' of 1904."
7. Brown, "Book Museum at Brussels," 338.
8. Otlet, as quoted (translated) in Rayward, La Fontaine, and Otlet, *Mundaneum*, 22.
9. Otlet, as quoted (translated) in Rayward, *Universe of Information*, 158.
10. Ibid.
11. Mazower, *Governing the World*, 105.
12. Leonards and Randeraad, "Building a Transnational Network of Social Reform in the 19th Century."
13. Otlet, as quoted (translated) in Rayward, *Universe of Information*, 161.
14. Otlet, as quoted (translated) in ibid., 159.
15. Otlet and La Fontaine, as quoted (translated) in ibid., 172.
16. Bourgeois, *Pour la société des nations*.
17. Otlet, as quoted (translated) in Rayward, *Universe of Information*, 172.
18. Otlet, "On a New Form of the Book," 91.
19. Ibid.
20. Otlet, as quoted in Levie, *L'Homme qui voulait classer le monde*, 98. Translation mine.
21. Otlet, as quoted (translated) in Rayward, *Universe of Information*, 161.
22. Otlet, "Preservation and International Diffusion of Thought," 208.
23. La Fontaine and Otlet, as quoted (translated) in Rayward, *Universe of Information*.
24. Otlet, as quoted in Levie, *L'Homme qui voulait classer le monde*, 110. Translation mine.
25. Rayward, *Universe of Information*, 145.
26. Otlet, as quoted in Levie, *L'Homme qui voulait classer le monde*, 111. Translation mine.
27. Ibid.

28. Rayward, *Universe of Information*, 187.
29. Dewey, as quoted in ibid.
30. Seymour, as quoted in ibid., 102.
31. Rayward, *Universe of Information*, 100–105.
32. Otlet, as quoted (translated) in ibid., 103.

Chapter 5

1. "Edinburgh's Camera Obscura and World of Illusions."
2. Smith, "Planning as Environmental Improvement," 99–133.
3. Geddes, *Cities in Evolution*, 320.
4. Defries, *Interpreter Geddes*, 95.
5. Welter, *Biopolis*, 17.
6. Geddes, *Classification of Statistics and Its Results*, 18–24.
7. Geddes, as quoted in Defries, *Interpreter Geddes*, 187–90.
8. Van Acker, *Universalism as Utopia*, 306.
9. Geddes, as quoted in Meller, *Patrick Geddes*, 43.
10. Geddes, "Index Museum," 66–67.
11. Geddes, quoted in Welter, *Biopolis*, 130.
12. Mumford, "Patrick Geddes, Victor Branford and Applied Sociology in England," 381.
13. Rayward, *Universe of Information*, 120–92.
14. Otlet, as quoted (translated) in ibid., 193.
15. La Fontaine, as quoted in Carnegie Endowment report, as quoted in ibid., 191.
16. *Visite d'Andrew Carnegie Au Palais Mondial-Mundaneum/1913 (BXLS).*
17. Carnegie, as quoted in Levie, *L'Homme qui voulait classer le monde*, 149.
18. Ibid.
19. Rayward, *Universe of Information*, 192.
20. Geddes, as quoted in Van Acker, "La remédiation de la connaissance encyclopédique," 181–82.
21. Chabard, "Towers and Globes," 107–10.
22. Otlet, "Union of International Associations," 116.

Chapter 6

1. O. Andersen, "Diary."
2. Hendrik Andersen, introduction to Henry James letters, in Hendrik Christian Andersen Papers, undated, Box 17.
3. Ibid.
4. Tóibín, *The Master*, 271.

5. James, *Roderick Hudson.*
6. Tóibín, *The Master,* 273.
7. O. Andersen, "Diary."
8. James, *Beloved Boy,* xii–xiii.
9. Tóibín, *The Master,* 273.
10. Hendrik Andersen, quoted in James, *Beloved Boy,* xviii.
11. Hendrik Andersen, letter to Olivia Cushing Andersen, in Hendrik Christian Andersen Papers, June 17, 1913.
12. Andersen and Hébrard, *Creation of a World Centre of Communication.*
13. Van Acker, *Universalism as Utopia,* 587.
14. Andersen and Hébrard, *Creation of a World Centre of Communication,* vii.
15. Ibid., 14.
16. O. Andersen, "Diary," July 6, 1911.
17. [Urbain Ledoux], unsigned copy of a letter to Hendrik Andersen, January 27, 1911, from the Mundaneum Archives.
18. Ledoux, quoted in Andersen, "Diary," September 4, 1911.
19. Ibid.
20. Ledoux, quoted in ibid., November 2, 1911.
21. Hendrik Andersen, quoted in ibid., May 7, 1912.
22. Otlet, letter to Hendrik Andersen, February 24, 1912.
23. Hendrik Andersen, letter to Paul Otlet and Henri La Fontaine, March 12, 1912, Mundaneum Archives.
24. Hendrik Andersen, letter to Olivia Cushing Andersen and Mama, July 28, 1912, Hendrik Christian Andersen Papers.
25. Hendrik Andersen, journal entry, August 28, 1912. Hendrik Christian Andersen Papers.
26. "Pays to Lecture at Mr. Zero's Tub."
27. Schmidt, *Restless Souls,* 211.
28. Hendrik Andersen, letter to Henry James, March 31, 1912, Hendrik Christian Andersen Papers.
29. Ibid., April 14, 1913.
30. James, *Beloved Boy,* 101–3.
31. Hendrik Andersen, as quoted in ibid., 132–33.
32. James, *Beloved Boy,* xxii.
33. O. Andersen, "Diary," September 6, 1913.
34. Ibid., November 23, 1913.
35. Ibid.
36. Ibid.

Chapter 7

1. Moe, "Presentation Speech."
2. Zuckerman, The Rape of Belgium, 6.

3. Von Bethmann-Hollweg, "A Published Interview Explaining the 'Scrap of Paper' Phrase by German Chancellor Theobald von Bethmann-Hollweg."
4. La Fontaine, as quoted in Levie, *L'Homme qui voulait classer le monde*, 161. Translation mine.
5. Zuckerman, *The Rape of Belgium*.
6. Kramer, *Dynamic of Destruction*, 6–11.
7. Casement, as quoted in Mitchell, *Casement*, 102.
8. Otlet, "Personal Correspondence," Dossier 645.
9. Otlet, *La fin de la guerre*, 1.
10. Ibid., 7.
11. Jones and Sherman, *League of Nations*, 44–52.
12. Woolf, *International Government*.
13. La Fontaine, *Great Solution*.
14. Otlet, *La fin de la guerre*, 45.
15. Ibid.
16. Ibid., 25.
17. Ibid.
18. Levie, *L'Homme qui voulait classer le monde*, 169.
19. Otlet, "Note for M. Durand, Prefect of Police," 130.
20. Rayward, *Universe of Information*, 208–9.
21. Levie, *L'Homme qui voulait classer le monde*, 167
22. Otlet, as quoted (translated) in Rayward, *Universe of Information*, 205.
23. De Broqueville, as quoted in Levie, *L'Homme qui voulait classer le monde*, 174. Translation mine.
24. Otlet, as quoted in ibid., 175. Translation mine.
25. Ibid., 171. Translation mine.
26. Ibid., 175–76. Translation mine.
27. Sayers, as quoted in ibid., 242–43.
28. Bishop, as quoted in ibid., 243.
29. Richardson, as quoted in ibid., 244.
30. Ibid., 245.
31. Hendrik Andersen, letter to Paul Otlet, May 28, 1917, Mundaneum Archives.
32. Van Acker, *Universalism as Utopia*, 45.
33. Hendrik Andersen, letter to Paul Otlet, September 29, 1918, Hendrik Christian Andersen Papers. Box 23.
34. Hendrik Andersen, letter to Paul Otlet, November 18, 1918.
35. Hendrik Andersen, as quoted in Levie, *L'Homme qui voulait classer le monde*, 180.
36. Rayward, *Universe of Information*, 208–9.
37. Otlet, as quoted in Levie, *L'Homme qui voulait classer le monde*, 182. Translation mine.
38. Otlet, as quoted in ibid., 181. Translation mine.
39. Ibid., 183. Translation mine.

40. Drummond, as quoted in Rayward, *Universe of Information*, 213.
41. Otlet, as quoted (translated) in ibid., 238.
42. Otlet, as quoted (translated) in ibid., 245.
43. Ibid., 241.
44. Du Bois, as quoted in Abrahams, "Congress in Perspective."
45. Abrahams, "Congress in Perspective."
46. Otlet, "Suggestions of Paul Otlet Regarding the Pan African Congress."
47. Otlet, *Monde*, 86.
48. *Neptune*, as quoted (translated) in Abrahams, "Congress in Perspective."
49. Du Bois quoted in Fauset, "Impressions of the Second Pan-African Congress," 13.
50. Rayward, *Universe of Information*, 250–55.
51. Ibid., 270.

Chapter 8

1. Godet, as quoted (translated) in Rayward, *Universe of Information*, 278.
2. Otlet, *Conférence des Associations Internationales*.
3. Originally spelled "Mondaneum" but later changed to "Mundaneum."
4. Otlet, as quoted (translated) in Otlet, *Conférence des Associations Internationales*, 281.
5. Rayward, *Universe of Information*, 284–85.
6. Godet, as quoted (translated) in ibid., 291.
7. Ibid., 276–98.
8. Andersen, letter to Paul Otlet, March 20, 1928. Mundaneum Archives.
9. Le Corbusier, *Towards a New Architecture*, 269.
10. Jencks, *Le Corbusier and the Continual Revolution in Architecture*, 215.
11. Otlet and Le Corbusier, *Mundaneum*.
12. Ibid.
13. Ibid.
14. Ibid., 38–39.
15. Van Acker, "Opening the Shrine of the Mundaneum," 801.
16. Henderson, "J. L. M. Lauweriks and K. P. C. de Bazel: Architecture and Philosophy," 1.
17. Van Acker, "Opening the Shrine of the Mundaneum," 797–98.
18. Van Acker, *Universalism as Utopia*, 643.
19. Otlet and Le Corbusier, *Mundaneum*.
20. Van Acker, *Universalism as Utopia*, 644.
21. Otlet and Le Corbusier, *Mundaneum*, 43.
22. Van den Heuvel, "Building Society, Constructing Knowledge, Weaving the Web," 129.
23. Otlet and Le Corbusier, *Mundaneum*, 25–26.
24. Hendrik Andersen, letter to Paul Otlet, June 4, 1929. Mundaneum Archives.

25. Levie, *L'Homme qui voulait classer le monde*, 238.
26. Otlet, *L'Education et les Instituts Du Palais Mondial (Mundaneum)*, 3.
27. Ibid., 24.
28. Henning, *Museums, Media and Cultural Theory*, 25.
29. Ibid., 31.
30. Otlet, "Encyclopedia Universalis Mundaneum Papers."
31. Otlet, "Chacun devenant son propre editeur."
32. Van Acker, "La remédiation de la connaissance encyclopédique."
33. Ibid.
34. Ibid.
35. Vossoughian, "Modern Museum in the Age of Its Mechanical Reproducibility," 241.
36. Neurath, as quoted (translated) in ibid., 242.
37. Ibid.
38. Vossoughian, *Otto Neurath*, 91.
39. Neurath, as quoted (translated) in ibid., 104.
40. Neurath and Neurath, *Empiricism and Sociology*, 218.
41. Neurath, as quoted (translated) in Van Acker, "Internationalist Utopias of Visual Education," 66–67.
42. Neurath, as quoted (translated) in Nikolow, "Gesellschaft Und Wirtschaft," 262.
43. Neurath and Neurath, *Empiricism and Sociology*, 219.
44. Vossoughian, "Language of the World Museum," 82–93.
45. Van Acker, *Universalism as Utopia*, 338.
46. Van Acker, "La remédiation de la connaissance encyclopédique."
47. Otlet, "Systematic Organization of Documentation and the Development of the International Institute of Bibliography," 117.
48. Otlet, as quoted (translated) in Rayward, *Universe of Information*, 347.
49. Otlet, quoted in ibid., 327.
50. Rayward, *Universe of Information*, 329.
51. Levie, *L'Homme qui voulait classer le monde*, 263.
52. Capart, as quoted in ibid., 265. Translation mine.
53. Otlet, quoted in Rayward, *Universe of Information*, 351.

Chapter 9

1. Ostwald, quoted in Hapke, "Roots of Mediating Information," 315.
2. Otlet, *Monde*, 390–91.
3. Krajewski, "Die Brücke."
4. Buckland, *Emanuel Goldberg and His Knowledge Machine*, 65.
5. Ostwald, "Biology of the Savant," 170.
6. Hapke, "Roots of Mediating Information," 317.

7. Buckland, "Wilhelm Ostwald and the Bridge."
8. Buckland, *Emanuel Goldberg and His Knowledge Machine*, 111.
9. Ibid., 148.
10. Ibid., 154–156.
11. "James W. Bryce Biography."
12. Van den Heuvel, "Dutch Connection."
13. "Documentation Congress Step toward Making 'World Brain.'"
14. Richards, *Scientific Information in Wartime*, 64.
15. Wells, "Today's Distress and Horrors Basically Intellectual," 229.
16. Wells, *World Brain*, 119.
17. Ibid., 121.
18. Ibid., 95.
19. Ibid., 100–105.
20. Ibid., 113.
21. Wells, as quoted in Rayward, "March of the Modern and the Reconstitution of the World's Knowledge Apparatus," 229.
22. Rayward, *Universe of Information*, 357.
23. Buckland, *Emanuel Goldberg and His Knowledge Machine*, 222.
24. Buckland, "Emanuel Goldberg, Electronic Document Retrieval, and Vannevar Bush's Memex."
25. Wells, *World Brain*, 86–87.
26. Teilhard de Chardin, "Hominization" in *The Vision of the Past*.
27. Borges, "Analytical Language of John Wilkins."
28. Ibid.
29. Otlet, "Science of Bibliography," 72–73.
30. Eco, *Search for the Perfect Language*, 259.

Chapter 10

1. A now-defunct classification; jellyfish are now considered part of the phylum Cnidaria.
2. Levie, *L'Homme qui voulait classer le monde*, 70.
3. Rayward, *Universe of Information*, 350.
4. Ibid., 353.
5. Otlet, *Traité de documentation*, [1].
6. Ibid., 291.
7. Ibid., 357.
8. Ibid., 370.
9. Ibid., 267–69.
10. Ibid., 216.
11. Ibid., 431.
12. Ibid., 43.

13. Ibid., 9.
14. Ibid., 254.
15. Ibid., 278.
16. Ibid., 374.
17. Day, *Modern Invention of Information*, 14.
18. Van Acker, *Universalism as Utopia*, 130–131.
19. Day, *Modern Invention of Information*, 19.
20. Otlet, "Preservation and International Diffusion of Thought," 208.
21. Otlet, *Traité de documentation*, 387–91.
22. Ibid.
23. Ibid., 428.
24. Ibid., 391.
25. Ibid., 423.
26. Ibid., 415.
27. Ibid., 422.
28. Ibid.
29. Otlet, as quoted (translated) in Van Acker, *Universalism as Utopia*, 222.
30. Ibid., 429.
31. Ibid., 427.
32. Otlet, unpublished drawing of Mundaneum, 1943; Mundaneum Archives.
33. Otlet, *Traité de documentation*, 430.
34. Otlet, *Monde*, xiii.
35. Ibid., 86.
36. Ibid., 109.
37. Ibid., 239–40.
38. Ibid., 101.
39. Ibid., 202.
40. Ibid.
41. Ibid., 382.
42. Ibid., 386.
43. Otlet, *Plan Belgique*, 146.
44. Otlet, *Traité de documentation*, 431.

Chapter 11

1. Licklider, "Memorandum for Members and Affiliates of the Intergalactic Computer Network."
2. Licklider, *Libraries of the Future*, 5–6.
3. Shera, as quoted in Van den Heuvel, "Multidimensional Classifications," 453.
4. Shera, as quoted in ibid., 454.
5. Leiner et al., "Brief History of the Internet."

6. Cerf, Dalal, and Sunshine, "Specification of Internet Transmission Control Program."
7. WWW Consortium, "General Overview."
8. Contemporary Web critics like Ted Nelson have pointed out that despite its foundational rhetoric about openness, the Web does indeed rely on hierarchical file and directory structures—like XML (which Nelson describes as a "hierarchy hamburger")—but those criticisms aside, most users experience the Web as a flat, open environment of seemingly unlimited access to information.
9. Weinberger, *Everything Is Miscellaneous.*
10. Bush, "As We May Think."
11. Although there is no evidence that Goldberg's invention directly influenced Bush's article, Bush later learned about Goldberg's invention in 1946.
12. Bush, "As We May Think."
13. Ibid.
14. Engelbart, quoted in Markoff, *What the Dormouse Said,* 66–67.
15. Markoff, *What the Dormouse Said,* 148–50.
16. Kesey, as quoted in ibid., 165.
17. Brand, "Spacewar: Fanatic Life and Symbolic Death among the Computer Bums."
18. Nelson, "Ted Nelson Specs."
19. Nelson, *Possiplex,* 100–101.
20. Ibid., 36.
21. Van den Heuvel, private correspondence.
22. Nelson, *Literary Machines 93.1,* 2.
23. Nelson, "File Structure for the Complex, the Changing and the Indeterminate," 134.
24. Nelson, *Computer Lib/Dream Machines.*
25. Nelson, "File Structure for the Complex, the Changing and the Indeterminate."
26. Ibid.
27. Nelson, *Literary Machines 93.1,* 1–23.
28. Ibid., 1–21.
29. Duyvis, "The UDC: What It Is and What It Is Not"; cited by Van den Heuvel in private correspondence.
30. Nelson, *Geeks Bearing Gifts,* 68.

Chapter 12

1. Berners-Lee, "Information Management."
2. Berners-Lee, "HyperText and CERN."
3. Berners-Lee, "The World Wide Web and the 'Web of Life.'"
4. Church, "One Light, Many Windows."
5. Berners-Lee, "The World Wide Web and the 'Web of Life.'"

6. Berners-Lee, "Keynote."
7. Berners-Lee, Hendler, and Lassila, "The Semantic Web."
8. Van den Heuvel, "Web 2.0 and the Semantic Web in Research from a Historical Perspective."
9. Van den Heuvel, private correspondence.
10. Van den Heuvel, "Web 2.0 and the Semantic Web in Research from a Historical Perspective."
11. Hillis, "Hillis Knowledge Web."
12. Wright, "Data Streaming 2.0."
13. Van den Heuvel, "Web 2.0 and the Semantic Web in Research from a Historical Perspective."
14. Shirky, "Ontology Is Overrated."
15. Weinberger, *Everything Is Miscellaneous*, 102.
16. Ibid., 100.
17. Ibid., 103.
18. Lee, "Citizendium Turns Five, but the Wikipedia Fork Is Dead in the Water."
19. Joslyn and Turchin, "Introduction to the 'Principia Cybernetica' Project."
20. Heylighen, "Conception of a Global Brain."
21. Kelly, "We Are the Web."
22. Lévy, *Collective Intelligence*, 13–15.
23. "Internet 2012 in Numbers."
24. Gelernter, "The End of the Web, Search, and Computer as We Know It."
25. Rayward, "Visions of Xanadu."
26. Otlet, *Traité de Documentation*, 431.
27. Hillis, "The Big Picture."
28. Kurzweil, "Law of Accelerating Returns."
29. Shirky, *Cognitive Surplus*.

Conclusion

1. Echikson, personal interview.
2. Dyson, "Turing's Cathedral."
3. Morris, "Henry James and an Eccentric Sculptor's Fantasies."
4. Mumford, introduction to Geddes and Tyrwhitt, *Patrick Geddes in India*, 9.
5. Brook, "Heirs Apparent," 17.
6. Le Corbusier, *Radiant City*, 263.
7. Friedman, *The World Is Flat*, 92.
8. Leonard, "Thanks for Nothing, 1 Percent!"

BIBLIOGRAPHY

Abrahams, Peter. "The Congress in Perspective." In *History of the Pan-African Congress*, edited by George Padmore, 1947. http://www.marxists.org/archive/padmore/1947/pan-african-congress/ch05.htm.

Andersen, Hendrik Christian. Hendrik Christian Andersen Papers, n.d. Manuscript Division. Library of Congress.

Andersen, Hendrik Christian, and Ernest M. Hébrard. *Creation of a World Centre of Communication*. Paris: Philippe Renouard, 1913.

Andersen, Olivia Cushing. "Diary," n.d. Manuscript Division. Library of Congress.

Bacon, Francis. *Sylva Sylvarum; or, A Natural History, in Ten Centuries*. Accessed March 31, 2013. http://www.sirbacon.org/sylvasylvarumpreface.htm.

Batchen, Geoffrey. "Electricity Made Visible." In *New Media, Old Media: A History and Theory Reader*, edited by Wendy Hui Kyong Chun and Thomas Keenan. New York: Routledge, 2006.

Berners-Lee, Tim. "HyperText and CERN." Accessed July 22, 2013. http://www.w3.org/Administration/HTandCERN.txt.

———. "Information Management: A Proposal," March 1989. http://www.w3.org/History/1989/proposal.html.

———. "Keynote." Presented at the WWW5 Conference, Paris, France, May 8, 1996. http://www5.wwwconference.org/fich_html/plenary-sessions.html.

———. "The World Wide Web and the 'Web of Life.'" Accessed November 15, 2012. http://www.w3.org/People/Berners-Lee/UU.html.

Berners-Lee, Tim, James Hendler, and Ora Lassila. "The Semantic Web: Scientific American." *Scientific American*, May 2001. http://www.sciam.com/article.cfm?articleID=00048144-10D2-1C70-84A9809EC588EF21&pageNumber=1&catID=2.

Blair, Ann. *Too Much to Know: Managing Scholarly Information before the Modern Age.* New Haven, Conn.: Yale University Press, 2010.

Borges, Jorge Luis. "The Analytical Language of John Wilkins." In *Other Inquisitions, 1937–1952.* Austin: University of Texas Press, 1964.

———. "The Library of Babel." In *Ficciones,* translated by Anthony Kerrigan. New York: Grove Press, 1962.

Bourgeois, Lèon. *Pour la société des nations.* Paris: Bibliothèque-Charpentier, 1910.

Boyd, J. P. *The Paris Exposition of 1900: A Vivid Descriptive View and Elaborate Scenic Presentation of the Site, Plan and Exhibits.* Philadelphia: P. W. Ziegler & Co., 1900.

Brand, Stewart. "Spacewar: Fanatic Life and Symbolic Death among the Computer Bums." *Rolling Stone,* December 7, 1972.

Brook, Daniel. "Heirs Apparent." *Harper's,* January 2013.

Brooker, P., S. Bru, A. Thacker, and C. Weikop. *The Oxford Critical and Cultural History of Modernist Magazines, Volume III: Europe 1880–1940.* Oxford: Oxford University Press, 2013.

Brown, James Duff. "The Book Museum at Brussels." *Library World* 10 (July 1907): 337–38.

Buckland, Michael. "Emanuel Goldberg, Electronic Document Retrieval, and Vannevar Bush's Memex." *Journal of the American Society for Information Science* 43, no. 4 (May 1992): 284–94.

———. "On the Cultural and Intellectual Context of European Documentation." In *European Modernism and the Information Society,* edited by W. Boyd Rayward, 45–57. Aldershot: Ashgate, 2008.

———. "Wilhelm Ostwald and the Bridge." Accessed October 24, 2012. http://people.ischool.berkeley.edu/~buckland/ponto.html.

Buckland, Michael Keeble. *Emanuel Goldberg and His Knowledge Machine: Information, Invention, and Political Forces.* Westport, Conn.: Libraries Unlimited, 2006.

Bush, Vannevar. "As We May Think." *Atlantic Monthly* 176 (July 1945): 641–49.

Cerf, Vinton, Yogen Dalal, and Carl Sunshine. "Specification of Internet Transmission Control Program." Network Working Group, December 1974. http://tools.ietf.org/html/rfc675.

Chabard, Pierre. "Towers and Globes: Architectural and Epistemological Differences between Patrick Geddes's Outlook Towers and Paul Otlet's Mundaneums." In *European Modernism and the Information Society,* edited by W. Boyd Rayward. Aldershot: Ashgate, 2008.

De Chardin, Teilhard. "Hominization and speciation." In The Vision of the Past: The Basis and Foundations of the Idea of Evolution. Translated by J.M. Cohen. New York: Harper and Row, 1966. 256–267.

Church, Forrest. "One Light, Many Windows," October 6, 2002. http://www.allsoulsnyc2.org/publications/sermons/fcsermons/one-light-many-windows.html.

Columbanus, Hibernus. *Monk's Rules*, translated by G. S. M. Walker. Second draft, revised and corrected. CELT: Corpus of Electronic Texts. Cork, Ireland: University College. Accessed June 28, 2013. http://www.ucc.ie/celt/online/T201052.html.

Day, Ronald E. *The Modern Invention of Information: Discourse, History, and Power*. Carbondale: Southern Illinois University Press, 2001.

Defries, Amelia. *The Interpreter Geddes, the Man and His Gospel*. London: Routledge, 1927.

Dewey, Melvil. *A Classification and Subject Index for Cataloguing and Arranging the Books and Pamphlets of a Library*. Amherst, Mass., 1876. http://www.gutenberg.org/files/12513/12513-h/12513-h.htm.

———. "[Editorial]." *Library Journal*, no. 1 (May 1877): 321–22.

Diderot, Denis. *Oeuvres de Denis Diderot: Prospectus de L'encyclopédie. Lettrex Au R.P. Berthier. Dictionnaire Encyclopédique: AB-CY, Volumes 1–2*. Paris: Belin, 1818.

"Documentation Congress Step toward Making 'World Brain.'" *Science News-Letter* 32, no. 861 (October 9, 1937): 228–29.

Duyvis, Donker. "The UDC: What It Is and What It Is Not." *Revue de La Documentation* 18, no. 2 (1951): 99–105.

Dyson, George. *Darwin among the Machines: The Evolution of Global Intelligence*. Reading, Mass.: Perseus, 1998.

———. "Turing's Cathedral." *Edge*, October 23, 2005. http://edge.org/conversation/turing-39s-cathedral.

Echikson, William. Personal interview, September 21, 2012.

Eco, Umberto. *The Search for the Perfect Language*. Oxford: Blackwell, 1997.

———. "Vegetal and Mineral Memory: The Future of Books." Bibliotheca Alexandrina, Alexandria, Egypt, November 1, 2003. http://www.bibalex.org/eminentlectures/lecturedetails_en.aspx?ID=23.

"Edinburgh's Camera Obscura and World of Illusions: History of the Attraction." Accessed August 12, 2013. http://camera-obscura.co.uk/camera_obscura/camera_history_attraction.asp.

Eisenstein, Elizabeth L. *The Printing Revolution in Early Modern Europe*. Cambridge: Cambridge University Press, 2012.

Emerson, Barbara. *Leopold II of the Belgians: King of Colonialism*. London: Weidenfeld and Nicolson, 1979.

Erasmus, D., and W. W. Barker. *The Adages of Erasmus*. Toronto: University of Toronto Press, 2001.

Fauset, Jessie. "Impressions of the Second Pan-African Congress," edited by W. E. B. Du Bois. *The Crisis: A Record of the Darker Races* 23, no. 133 (November 1921): 12–18.

Febvre, Lucien, and Henri-Jean Martin. *The Coming of the Book: The Impact of Printing 1450–1800*. London: Verso, 1997.

Foxe, John. *The Unabridged Acts and Monuments Online*. 1570th ed. Sheffield: HRI Online Publications, 2011. http//www.johnfoxe.org.

Friedman, Thomas L. *The World Is Flat: A Brief History of the Twenty-First Century*. New York: Farrar, Straus and Giroux, 2007.

Geddes, Patrick. *Cities in Evolution: An Introduction to the Town Planning Movement and to the Study of Civics*. London: Williams & Norgate, 1915.

———. *Civics: As Applied Sociology*. 1904.

———. *The Classification of Statistics and Its Results*. Vol. 11. Proceedings of the Royal Society of Edinburgh. Edinburgh: A. & C. Black, 1881.

———. "The Index Museum: Chapters from an Unpublished Manuscript." *Assemblage*, no. 10 (December 1989): 65–69.

Geddes, P., and J. Tyrwhitt. *Patrick Geddes in India*. London: Lund Humphries, 1947.

Gelernter, David. "The End of the Web, Search, and Computer as We Know It." *Wired*, February 1, 2013. http://www.wired.com/opinion/2013/02/the-end-of-the-web-computers-and-search-as-we-know-it/.

Gessner, Conrad. *Bibliotheca Universalis Und Appendix*. Milliaria, 5 v. 1. O. Zeller, 1966.

Greenblatt, Stephen. *The Swerve: How the World Became Modern*. New York: Norton, 2012.

Hapke, Thomas. "Roots of Mediating Information: Aspects of the German Information Movement." In *European Modernism and the Information Society*, edited by W. Boyd Rayward, 314–27. Aldershot: Ashgate, 2008.

Henderson, Susan R. "J. L. M. Lauweriks and K. P. C. de Bazel: Architecture and Philosophy." *Architronics* 7, no. 2 (1998): 1–15.

Henning, Michelle. *Museums, Media and Cultural Theory* (Issues in Cultural and Media Studies). Maidenhead, Berkshire: Open University Press, 2005.

Henri La Fontaine, Prix Nobel de La Paix en 1913: Un Belge Épris de Justice. Bruxelles: Éditions Racine, 2012.

Heylighen, Francis. "Conception of a Global Brain: An Historical Review." In *From Big Bang to Global Civilization: A Big History Anthology*, edited by Barry Rodrigue, Leonid Grinin, and Audrey Korotayev. Berkeley: University of California Press, forthcoming. http://pespmc1.vub.ac.be/Papers/GB-Conceptions-Rodrigue.pdf.

Hillis, W. Daniel. "The Big Picture." *Wired*, January 1998. http://www.wired.com/wired/archive/6.01/hillis.html.

———. "The Hillis Knowledge Web." *Edge*, July 18, 2010. http://edge.org/conversation/the-hillis-knowledge-web.

Hochschild, Adam. *King Leopold's Ghost: A Story of Greed, Terror, and Heroism in Colonial Africa*. Boston: Houghton Mifflin, 1999.

Hunter, E. J. *Classification Made Simple: An Introduction to Knowledge Organisation and Information Retrieval*. Aldershot: Ashgate, 2009.

"Industrial History | Belgium." European Route of Industrial Heritage. Accessed July 27, 2013. http://www.erih.net/topmenu/about-erih.html.

"Internet 2012 in Numbers." Pingdom, January 16, 2013. http://royal.pingdom.com/2013/01/16/internet-2012-in-numbers/.

Jacobs, Jane. *The Death and Life of Great American Cities*. New York: Vintage, 1992.

James, Henry. *Beloved Boy: Letters to Hendrik C. Andersen, 1899–1915*. Charlottesville: University of Virginia Press, 2004.

———. *Roderick Hudson*. Boston: Houghton Mifflin, 1917.

"James W. Bryce Biography." IBM. Accessed July 18, 2013. http://www-03.ibm.com/ibm/history/exhibits/markI/2413JB01.html.

Jencks, Charles. *Le Corbusier and the Continual Revolution in Architecture*. New York: Monacelli, 2000.

Jones, Robert, and Sherman, S. S. *The League of Nations: From Idea to Reality*. London: Sir Isaac Pitman & Sons, Ltd., 1927.

Joslyn, Cliff, and Valentin Turchin. "Introduction to the 'Principia Cybernetica' Project." Accessed December 21, 2012. http://pespmc1.vub.ac.be/FPUBINT.html.

Kelly, Kevin. "We Are the Web." *Wired*, August 2005. Available at http://www.wired.com/wired/archive/13.08/tech.html.

Krajewski, Markus. "Die Brücke: A German Contemporary of the Institut International de Bibliographie." *Cahiers de la Documentation* 2 (June 2012).

———. *Paper Machines: About Cards and Catalogs, 1548–1929*. Cambridge, Mass.: MIT Press, 2011.

Kramer, Alan. *Dynamic of Destruction: Culture and Mass Killing in the First World War*. Oxford: Oxford University Press, 2007.

Kurtz, M. J. *America and the Return of Nazi Contraband: The Recovery of Europe's Cultural Treasures*. Cambridge: Cambridge University Press, 2006.

"Kurzbiographien/Dictionnaire Biographique." Institut d'Histoire du Temps Présent—IHTP. Accessed June 28, 2013. http://www.ihtp.cnrs.fr/prefets/kurzbiographien_cv.html.

Kurzweil, Ray. "The Law of Accelerating Returns," March 7, 2001. http://www.kurzweilai.net/the-law-of-accelerating-returns.

Laaff, Meike. "Internet Visionary Paul Otlet: Networked Knowledge, Decades before Google." *Spiegel Online*, July 22, 2011. http://www.spiegel.de/international/world/internet-visionary-paul-otlet-networked-knowledge-decades-before-google-a-775951.html.

La Fontaine, Henri. *The Great Solution; Magnissima Charta*. Boston: World Peace Foundation, 1916. http://archive.org/details/greatsolutionma01fontgoog.

Le Corbusier. *The Radiant City: Elements of a Doctrine of Urbanism to Be Used as the Basis of Our Machine-Age Civilization*. New York: Orion Press, n.d.

———. *Towards a New Architecture*, translated by Frederick Etchells. London: Butterworth, 1989.

Lee, Thomas B. "Citizendium Turns Five, but the Wikipedia Fork Is Dead in the Water." *Ars Technica*, October 27, 2011.

Leiner, Barry M., et al. "Brief History of the Internet." Internet Society. Accessed July 19, 2013. http://www.internetsociety.org/internet/what-internet/history-internet/brief-history-internet.

Leonard, Andrew. "Thanks for Nothing, 1 Percent! A Selfish Silicon Valley Must Learn from History." *Salon*, August 9, 2013. http://www.salon.com/2013/08/09/thanks_for_nothing_1_percent_a_selfish_silicon_valley_must_learn_from_history/.

Leonards, Chris, and Nico Randeraad. "Building a Transnational Network of Social Reform in the 19th Century." In *Transnational Networks of Experts and Organizations (1850–1930)*. New York: Berghahn, forthcoming.

Levie, Françoise. *L'Homme qui voulait classer le monde: Paul Otlet et le Mundaneum*. Bruxelles: Les Impressions Nouvelles, 2006.

Lévy, Pierre. *Collective Intelligence: Mankind's Emerging World in Cyberspace*. Cambridge, Mass.: Perseus, 1997.

Library of Congress, "Fascinating Facts." Accessed March 31, 2013. http://www.loc.gov/about/facts.html.

Licklider, J. C. R. *Libraries of the Future*. Cambridge, Mass.: MIT Press, 1965.

———. "Memorandum for Members and Affiliates of the Intergalactic Computer Network." Kurzweil AI Net (republished December 11, 2011), April 23, 1963. http://www.kurzweilai.net/memorandum-for-members-and-affiliates-of-the-intergalactic-computer-network.

Markoff, John. *What the Dormouse Said: How the Sixties Counterculture Shaped the Personal Computer Industry*. New York: Penguin, 2006.

Mazower, Mark. *Governing the World: The History of an Idea*. New York: Penguin, 2012.

Meller, Helen. *Patrick Geddes: Social Evolutionist and City Planner* (Routledge Geography, Environment, and Planning Series). New York: Taylor & Francis, 1994.

Mitchell, Angus. *Casement (Life & Times)*. London: Haus, 2003.

Moe, Ragnvald. "Presentation Speech." December 10, 1913. http://www.nobelprize.org/nobel_prizes/peace/laureates/1913/press.html.

Morris, Roderick Conway. "Henry James and an Eccentric Sculptor's Fantasies." *New York Times*, June 3, 2000. http://www.nytimes.com/2000/06/03/style/03iht-james.t.html.

Mumford, Lewis. "Patrick Geddes, Victor Branford and Applied Sociology in England: The Social Survey, Regionalism and Urban Planning." In *An Introduction to the History of Sociology*, edited by H. E. Barnes. Abridged ed., 1966. Chicago: University of Chicago Press, 1948.

Nelson, Theodor H. *Computer Lib/Dream Machines*. Chicago: Nelson, c. 1974.

———. "A File Structure for the Complex, the Changing and the Indeterminate." In *The New Media Reader*, edited by Noah Wardrip-Fruin and Nick Montfort, 134–48. Cambridge, Mass.: MIT Press, 2003.

———. *Geeks Bearing Gifts: V.1.1: How the Computer World Got This Way*. Sausalito, Calif.: Mindful Press, 2008.

———. *Literary Machines 93.1*. Sausalito, Calif.: Mindful Press, 1993.

———. *Possiplex: Movies, Intellect, Creative Control, My Computer Life and the Fight for Civilization: An Autobiography of Ted Nelson*. Sausalito, Calif.: Mindful Press, 2010.

———. "Ted Nelson Specs," n.d. http://hyperland.com/mlawLeast.html.
Neurath, Otto, and Marie Neurath. *Empiricism and Sociology: With a Selection of Biographical and Autobiographical Sketches*. Dordrecht: Reidel, 1973.
Nikolow, Sybilla. "Gesellschaft Und Wirtschaft: An Encyclopedia in Otto Neurath's Pictorial Statistics from 1930." In *European Modernism and the Information Society*, edited by W. Boyd Rayward, 257–278. Aldershot: Ashgate, 2008.
Ostwald, Wilhelm. "Biology of the Savant." *Scientific American Supplement* 72 (1911): 169–71.
Otlet, Paul. *L'Afrique aux noirs*. Bruxelles: Ferdinand Larcier, 1888.
———. "Chacun devenant son propre editeur," n.d. Encyclopedia Universalis Mundaneum papers (unpublished), Mundaneum.
———. *Conférence des Associations Internationales—Programme, Rapport Général*. Publication no. 113. Bruxelles: UIA, 1924.
———. *L'Education et les Instituts du Palais Mondial (Mundaneum)*. Publication no. 121. Bruxelles: UIA, 1926.
———. "Encyclopedia Universalis Mundaneum Papers." Unpublished, 1926–1944. Mundaneum.
———. *La fin de la guerre*. La Haye: Martinus Nijhoff, 1914.
———. "Journal Intime," 1882–1892. Mundaneum.
———. *Monde: Essai d'universalisme*. Bruxelles: Editiones Mundaneum, 1935.
———. "Note for M. Durand, Prefect of Police." In *Selected Essays of Paul Otlet*, translated by W. Boyd Rayward, 130–35. Amsterdam: Elsevier, 1990.
———. "On a New Form of the Book: The Microphotographic Book." In *Selected Essays of Paul Otlet*, translated by W. Boyd Rayward, 87–95. Amsterdam: Elsevier, 1990.
———. "Personal Correspondence," n.d. Mundaneum.
———. *Plan Belgique*. Bruxelles: Editiones Mundaneum, 1935.
———. "The Preservation and International Diffusion of Thought: The Microphotic Book." In *International Organisation and Dissemination of Knowledge: Selected Essays of Paul Otlet*, translated by W. Boyd Rayward, 204–10. Amsterdam: Elsevier, 1990.
———. "Science of Bibliography." In *International Organisation and Dissemination of Knowledge: Selected Essays of Paul Otlet*, translated by W. Boyd Rayward, 71–86. Amsterdam: Elsevier, 1990.
———. "Something about Bibliography." In *International Organisation and Dissemination of Knowledge: Selected Essays of Paul Otlet*, translated by W. Boyd Rayward, 11–24. Amsterdam: Elsevier, 1990.
———. "Suggestions of Paul Otlet Regarding the Pan African Congress," June 4, 1921. MS 312. W. E. B. Du Bois Papers, Special Collections and University Archives, University of Massachusetts Amherst Libraries. http://oubliette.library.umass.edu/view/full/mums312-b019-i033.
———. "Sur la structure des nombres classificateurs." *IIB Bulletin*, no. 1 (1895): 230–43.

———. "The Systematic Organization of Documentation and the Development of the International Institute of Bibliography." In *International Organisation and Dissemination of Knowledge: Selected Essays of Paul Otlet*, translated by W. Boyd Rayward, 105–111. Amsterdam: Elsevier, 1990.

———. *Traité de documentation*. Bruxelles: Editiones Mundaneum—Palais Mondial, 1934.

———. "Transformations of the Bibliographical Apparatus of the Sciences." In *International Organisation and Dissemination of Knowledge: Selected Essays of Paul Otlet*, translated by W. Boyd Rayward, 148–56. Amsterdam: Elsevier, 1990.

———. "The Union of International Associations: A World Centre." In *International Organisation and Dissemination of Knowledge: Selected Essays of Paul Otlet*, translated by W. Boyd Rayward, 112–29. Amsterdam: Elsevier, 1990.

Otlet, Paul, and Le Corbusier. *Mundaneum*. L'Union des Associations Internationales 128. Bruxelles: Palais Mondial, 1928.

"Pays to Lecture at Mr. Zero's Tub; Mirza Ahmad Sohrab Discusses Baha-U-Llah and Club Listens for $50." *New York Times*, November 17, 1930.

Postman, Neil. *Amusing Ourselves to Death: Public Discourse in the Age of Show Business*. New York: Penguin, 2006.

Pritchard, Stephen. "Risk of Information Overload That Threatens Business Growth." *Financial Times*, November 7, 2012. Special section on "The Connected Business."

Rayward, W. Boyd. "The Case of Paul Otlet, Pioneer of Information Science, Internationalist, Visionary: Reflections on Biography." *Journal of Librarianship and Information Science* 23 (1991): 135–45. Available at http://people.lis.illinois.edu/~wrayward/otlet/PAUL_OTLET_REFLECTIONS_ON_BIOG.HTM.

———. "Knowledge Organisation and a New World Polity: The Rise and Fall and Rise of the Ideas of Paul Otlet." *Transnational Associations* 1–2 (2003): 4–15.

———. "The March of the Modern and the Reconstitution of the World's Knowledge Apparatus: H. G. Wells, Encyclopedism and the World Brain." In *European Modernism and the Information Society*, edited by W. Boyd Rayward. Aldershot: Ashgate, 2007.

———. "Paul Otlet. Encyclopédiste, Internationaliste, Belge." In *Paul Otlet*. Brussels: Les Impressions Nouvelles, 2010.

———. *The Universe of Information: The Work of Paul Otlet for Documentation and International Organization*. FIT520. Moscow: VINITI for the International Federation for Documentation, 1975. http://hdl.handle.net/2142/651.

———. "Visions of Xanadu: Paul Otlet (1868–1944) and Hypertext." *Journal of the American Society of Information Science* 45 (1994): 235–50.

Rayward, W. Boyd, Henri La Fontaine, and Paul Otlet. *Mundaneum: Archives of Knowledge*. Urbana-Champaign: Graduate School of Library and Information Science, University of Illinois at Urbana-Champaign, 2010. Available at https://www.ideals.illinois.edu/handle/2142/15431.

Reef, Catherine. *Walt Whitman*. New York: Clarion, 1995.

Richards, Pamela Spence. *Scientific Information in Wartime*. Westport, Conn.: Greenwood Press, 1994.

Robida, Albert. *The Twentieth Century* (Wesleyan Early Classics of Science Fiction Series). Middletown, Conn: Wesleyan University Press, 2004.

Ryan, Timothy James. "About ESW and the Holocaust Museum." Accessed June 22, 2012. http://miresperanto.narod.ru/en/articles/holocaust.htm.

Schmidt, Leigh Eric. *Restless Souls: The Making of American Spirituality*. San Francisco: HarperSanFrancisco, 2005.

Shirky, Clay. *Cognitive Surplus: How Technology Makes Consumers into Collaborators*. New York: Penguin, 2011.

———. "Ontology Is Overrated." Accessed December 21, 2012. http://www.shirky.com/writings/ontology_overrated.html.

Simon, Gerd. Chronologie Krüss, Hugo Andres. Accessed September 11, 2013. http://homepages.uni-tuebingen.de/gerd.simon/ChrKruess.pdf

Sklair, Leslie. *The Sociology of Progress*. Abingdon: Routledge, 1970.

Smith, P. J. "Planning as Environmental Improvement: Slum Clearance in Victorian Edinburgh." In *Planning and the Environment in the Modern World, Vol. 1: The Rise of Modern Urban Planning 1800–1914*, edited by A. Sutcliffe, 99–133. London: Mansell, 1980.

Standage, Tom. *The Victorian Internet: The Remarkable Story of the Telegraph and the Nineteenth Century On-line Pioneers*. New York: Walker and Co., 1998.

Steenson, Molly Wright. "Interfacing with the Subterranean." *Cabinet*, no. 41 (Spring 2011): 82–86.

Taylor, Frederick Winslow. *The Principles of Scientific Management*. New York: Norton, 1967.

Tóibín, Colm. *The Master: A Novel*. New York: Scribner, 2005.

Twain, Mark. "From the 'London Times' of 1904." In *The Man That Corrupted Hadleyburg and Other Stories and Essays*, by Mark Twain, 128–146. New York: Harper & Brothers, 1900.

———. *King Leopold's Soliloquy*. Boston: P. R. Warren Co., 1905. http://msuweb.montclair.edu/~furrg/i2l/kls.html.

"UDC Structure & Tables." UDC Consortium. Accessed August 17, 2013. http://www.udcc.org/index.php/site/page?view=about_structure.

Unknown. "Einsatzstab Reichsleiter Rosenberg [papers]." Einsatzstab Reichsleiter Rosenberg, State Archives of the Ukraine. Accessed June 28, 2013. http://err.tsdavo.org.ua/.

U.S. Commission to the Paris Exposition, 1900, and F. W. Peck. *Report of the Commissioner-General for the United States to the International Universal Exposition, Paris, 1900. February 28, 1901*. v. 6. Govt. print. off., 1901.

Van Acker, Wouter. "Internationalist Utopias of Visual Education." *Perspectives on Science* 19 (2011): 32–80.

———. "Opening the Shrine of the Mundaneum. The Positivist Spirit in the Architecture of Le Corbusier and His Belgian 'Idolators.'" In *Proceedings of the Society of Architectural Historians, Australia and New Zealand: 30, Open*, edited by Alexandra Brown and Andrew Leach. Gold Coast, Qld: SAHANZ, 2013, 2: 791–805.

———. "La remédiation de la connaissance encyclopédique." In *Paul Otlet*, edited by Jacques Gillen. 177–98. Brussels: Belgium Les Impressions nouvelles, 2010.

———. *Universalism as Utopia*. Zelzate: University Press, 2011.

Van den Heuvel, Charles. "Building Society, Constructing Knowledge, Weaving the Web. Otlet's Visualizations of a Global Information Society and His Concept of a Universal Civilization." In *European Modernism and the Information Society*, edited by W. Boyd Rayward, 127–53. Aldershot: Ashgate, 2007.

———. "The Dutch Connection: Donker Duyvis and Perceptions of American and European Decimal Classification Systems in the First Half of the Twentieth Century." In *International Perspectives on the History of Information Science and Technology*, edited by Toni Carbo and Trudi Bellardo Hahn, 174–86. Medford, N.J.: Information Today, 2012.

———. "Multidimensional Classifications: Past and Future Conceptualizations and Visualizations." *Knowledge Organization* 39, no. 6 (2012): 446–60.

———. "Paul Otlet et les versions historiques de la genèse du World Wide Web, du Web Semantique et du Web 2.0". In *Paul Otlet et La Bibliologie, Fondateur du Mundaneum (1868–1944). Architecte du savoir, artisan de paix*, 159–175. Bruxelles: Les Impressions Nouvelles, 2010.

———. "Web 2.0 and the Semantic Web in Research from a Historical Perspective. The Designs of Paul Otlet (1868–1944) for Telecommunication and Machine Readable Documentation to Organize Research and Society." *Knowledge Organization* 36, no. 4 (2009): 214–26.

Visite d'Andrew Carnegie au Palais Mondial-Mundaneum/1913 (BXLS). Le Mundaneum, 1913. http://www.youtube.com/watch?v=Q2T7mk16zqs&list=PLjM-HaWxVFdUi4-133HuZ1dllZAvJ-3Nu&index=9.

Von Bethmann-Hollweg, Chancellor Theobald. "A Published Interview Explaining the 'Scrap of Paper' Phrase by German Chancellor Theobald von Bethmann-Hollweg." Accessed July 13, 2013. http://www.firstworldwar.com/source/scrapofpaper2.htm.

Vossoughian, Nader. "The Language of the World Museum: Otto Neurath, Paul Otlet, Le Corbusier." *Transnational Associations*, nos. 1–2 (2003): 82–93.

———. "The Modern Museum in the Age of Its Mechanical Reproducibility." In *European Modernism and the Information Society*, edited by W. Boyd Rayward, 241–55. Aldershot: Ashgate, 2008.

———. *Otto Neurath: The Language of the Global Polis*. Rotterdam: Nai Publishers, 2011.

Wall Randell, Sarah. "Dr. Faustus and the Printer's Devil." *SEL Studies in English Literature, 1500–1900* 48, no. 2 (Spring 2008): 259–81.

Watson, Peter. *Ideas: A History of Thought and Invention, from Fire to Freud.* New York: HarperCollins, 2005.

Weinberger, David. *Everything Is Miscellaneous: The Power of the New Digital Disorder.* New York: Times Books, 2007.

Wells, H. G. "Today's Distress and Horrors Basically Intellectual: Wells." *Science News-Letter* 32, no. 861 (October 9, 1937): 229.

———. *World Brain.* London: Adamantine Press, 1994.

Welter, Volker. *Biopolis: Patrick Geddes and the City of Life.* Cambridge, Mass.: MIT Press, 2002.

Whitman, Walt. *Leaves of Grass.* Project Gutenberg, 2008. http://www.gutenberg.org/files/1322/1322-h/1322-h.htm.

Wiegand, Wayne A. *Irrepressible Reformer: A Biography of Melvil Dewey.* Chicago: American Library Association, 1996.

Williams, Rosalind H. *Notes on the Underground: An Essay on Technology, Society, and the Imagination.* Cambridge, Mass.: MIT Press, 2008.

Woolf, Leonard. *International Government.* New York: Brentano's, 1916.

World of Mathematics. Detroit, Mich.: Gale Group, 2001.

Wright, Alex. "Data Streaming 2.0." *Communications of the ACM* 53, no. 4 (April 2010): 13–14. doi:10.1145/1721654.1721661.

———. *Glut: Mastering Information through the Ages.* Ithaca, N.Y.: Cornell University Press, 2008.

———. "The Web Time Forgot." *New York Times,* June 17, 2008, F1.

WWW Consortium. "General Overview," November 3, 1992. http://www.w3.org/History/19921103-hypertext/hypertext/DataSources/Top.html.

Zuckerman, Larry. *The Rape of Belgium: The Untold Story of World War I.* New York: New York University Press, 2004.

INDEX

Illustrations are indicated by italic page numbers.